Other Titles in This Serie

(Continued in the back of this publication)

Dynamics in Several
Complex Variables

Conference Board of the Mathematical Sciences

CBMS

Regional Conference Series in Mathematics

Number 87

Dynamics in Several Complex Variables

John Erik Fornæss

Published for the
Conference Board of the Mathematical Sciences
by the
American Mathematical Society
Providence, Rhode Island
with support from the
National Science Foundation

Expository Lectures
from the NSF-CBMS Regional Conference
held at SUNY at Albany, Albany, New York
June 12–17, 1994

Research partially supported by
National Science Foundation Grant DMS 9204097

1991 *Mathematics Subject Classification.* Primary 32H50;
Secondary 32H20, 58F23, 58F25.

Library of Congress Cataloging-in-Publication Data
Dynamics in several complex variables / [edited by] John Erik Fornæss.
    p.   cm. — (Regional conference series in mathematics, ISSN 0160-7642; no. 87)
    "Expository lectures from the NSF-CBMS regional conference held at SUNY at Albany, Albany, New York, June 12–17, 1994"—T.p. verso.
    Includes bibliographical references.
    ISBN 0-8218-0317-4 (alk. paper)
    1. Differentiable dynamical systems—Congresses. 2. Holomorphic functions—Congresses.
3. Holomorphic mappings—Congresses. 4. Iterative methods (Mathematics)—Congresses.
5. Functions of several complex variables—Congresses. I. Fornæss, John Erik. II. Series.
QA1.R33 no. 87
[QA614.8]
510 s—dc20
[515'.94]   95-44418
                                                                                        CIP

CONTENTS

CHAPTER 1

Introduction

The purpose of this lecture series was to introduce the audience to the literature on complex dynamics in higher dimension. This CBMS lecture series was held in Albany, New York, june 1994. Some of the lectures were up-dated versions of earlier lectures given in Montreal 1993, jointly with Nessim Sibony ([**FS8**]). The Montreal lectures were more based on pluripotential theory, while the present lectures avoided pluripotential theory completely. We hope to give an alternative expansion of the Montreal lectures, basing complex dynamics in higher dimension systematically on pluripotential theory.

What we are trying to do with these notes is to provide an easy to read introduction to the field, an introduction which motivates the topics. Moreover, the monograph should point the reader towards the technically more advanced literature. It is my feeling that mathematicians can read arbitrarily complicated material once they are motivated.

We will start our introduction by choosing a basic problem which everybody has seen before. The investigation of this problem will lead us naturally to studying Complex Dynamics in Higher Dimension. In order to understand the problem better, we are naturally lead to some dynamical questions. These dynamical questions are the ones which have been studied in the literature.

Our basic problem is to find roots of equations. This is not just a pedagogical trick because this was already a motivating problem for complex dynamics in the last century. A main tool for finding roots of equations is Newton's method. The idea of Newton's method is to guess a root and use this guess to find a better guess.

The first one to study complex dynamics was **Schröder**, 1870, 1871. Although Newton's method can, in its simplest form, be traced back to the Babylonians, he was the first to apply it to study **complex** roots of holomorphic polynomials of one complex variable. (See ([**Sc1**]) and ([**Sc2**]).) **We take as our starting point the problem to describe those initial guesses which lead to a root.**

To study this problem, let us introduce some notation. We can work in any dimension, but for simplicity we will mostly restrict our discussions to two complex variables. Moreover, we will restrict our discussion to roots of polynomial equations.

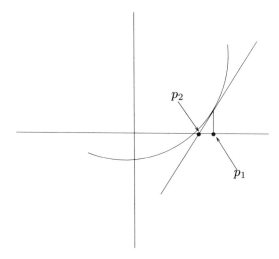

FIGURE 1.1. Newton's Method

Suppose we are given two polynomials of two complex variables. We are interested in their common roots.

$$\left\{ \begin{array}{ccc} A(z,w) & = & 0 \\ B(z,w) & = & 0 \end{array} \right\}.$$

Then the pair of polynomials can be considered as a map $R : \mathbb{C}^2 \mapsto \mathbb{C}^2$, $R = (A, B)$. The well known Newton's method then consists of guessing a root p_1 close to an actual root and getting hopefully an even closer point p_2 to the root from the formula

$$p_2 = p_1 - (R'(p_1))^{-1} R(p_1)$$

See figure 1.1. Hence we should obtain a root of $R = 0$ after infinitely many **iterations** of the map

$$F(p) = p - [(R')^{-1} R](p).$$

We get a sequence $\{p_n\}$,

$$p_{n+1} = F(p_n)$$

For some initial values p_1 the sequence $\{p_n\}$ does not converge to any root. So we come back to our initial problem again; **find the set of points p_1 for which $\{p_n\}$ converges to some root.**

As an example, we consider the equations:

$$\left\{ \begin{array}{cccc} A(z,w) & = & \frac{1}{2}z - \frac{\varepsilon}{2}w - \frac{1}{2}z^2 & = & 0 \\ B(z,w) & = & \frac{1}{2}w - \frac{\varepsilon}{2}z & = & 0 \end{array} \right.$$

Obviously $(0,0)$ is a common root. We will see how Newton's method works for this case. To simplify, one can first replace R' in the formula for Newton's method by the constant matrix

$$A = \begin{bmatrix} \frac{1}{2} & 0 \\ 0 & \frac{1}{2} \end{bmatrix},$$

if we assume that ε is small. We see that

$$F(z, w) = (z, w) - A^{-1}R = (z^2 + \varepsilon w, \varepsilon z)$$

Such maps are usually called **complex Hénon maps**. Newton's map F has a fixed point at the root, $(0,0)$. This point is an attracting fixed point of F. This means that the eigenvalues of the 2×2 matrix $F'(0,0)$ are smaller than one, which implies that all points p_1 in a small neighborhood of $(0,0)$ are attracted to $(0,0)$, $F^n(p_1) \rightarrow (0,0)$. The largest such set is an open subset of $\mathbb{C}^2$ and is called the **basin of attraction** of $(0,0)$. This set consists of all initial guesses giving that root.

In this case the basin of attraction is quite large, holomorphically equivalent to the whole space $\mathbb{C}^2$, nevertheless it is also quite small. So both the set of initial guesses giving the root and initial guesses not giving the root are quite large. A more precise description of these sets was only obtained in the last few years. See figure 1.2. Nevertheless, this kind of sets was studied already in the 20's and 30's by Fatou ([**Fa1**]) and Bieberbach ([**Bi**]). They are called Fatou-Bieberbach domains. See Rosay and Rudin, [**RR**], for many results on Fatou- Bieberbach domains and see Stensönes, [**S**], for a Fatou- Bieberbach domain with $\mathcal{C}^\infty$ boundary.

In the case of Hénon maps R, the iterates $\{F^n\}$ of Newton's method is a **normal family** on the basin of attraction of $(0,0)$. In general, when we have a map F, we call the largest open set where $\{F^n\}$ is a normal family the **Fatou set**. Moreover, we call the complement of the Fatou set the **Julia set**. So the Julia set is always closed.

In order to understand Newton's method, a main problem then is to describe the Fatou set and the Julia set of a map F. Since points on the Julia set are never roots of F, we would like to know whether the Julia set is small. In addition, some components of the Fatou set might not consist of initial guesses of any root. **Hence we are led to ask whether the Julia set has zero volume. Also we are led to ask what are the possible kinds of Fatou components.**

Next let us discuss the behaviour of Newton's Method on the Julia set of F. It is known that polynomial maps on $\mathbb{C}$ are always **chaotic** on their Julia sets. Hence we are led to ask whether F is chaotic on its Julia set as well.

A continuous map $F : K \rightarrow K$ on a compact metric space (K, d) is chaotic ([**De**]) if

i.- F is **sensitive to initial conditions**. This means that there exists a positive δ so that for any $x \in K$ and any positive ε, there is a point $y \in K$ closer to x than ε and an integer $n \geq 1$ so that $d(F^n(x), F^n(y)) > \delta$. See Figure 1.3.

ii.- F is **topologically transitive**. This means that whenever $x, y \in K$ and δ is a positive number, then there is a point $z \in K$ and an integer $n \geq 1$ so that both $d(x, z)$ and $d(F^n(z), y)$ are smaller than δ. See figure 1.4.

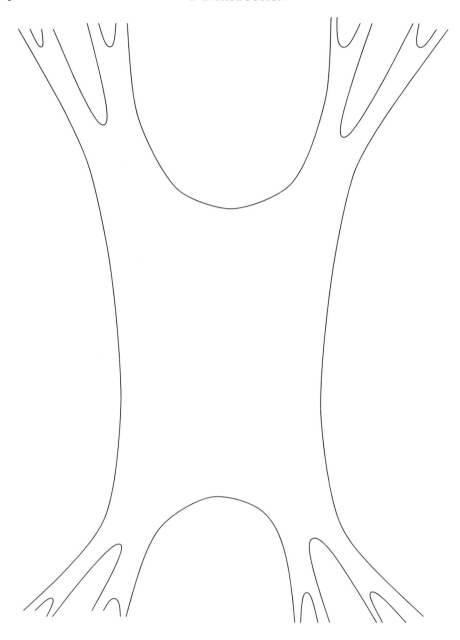

FIGURE 1.2. Fatou-Bieberbach Domain

iii.- The periodic orbits of F are dense in K.

It turns out that in dimensions above 1, the map F is not chaotic on the whole Julia set. For this reason it is more natural to study the **nonwandering** part of the Julia set. Suppose that F is a continuous self-map of some manifold M. Then a point $p \in M$ is nonwandering if given any neighborhood U of $p \in M$, there is an integer $n \geq 1$ so that $f^n(U) \cap U \neq \emptyset$. If not, p is **wandering.** See figure 1.5.

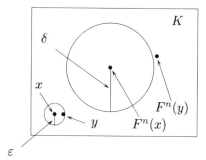

FIGURE 1.3. Sensitivity to initial conditions

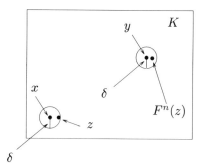

FIGURE 1.4. Topological transitivity

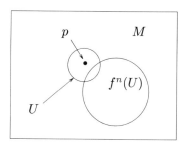

FIGURE 1.5. Nonwandering points

Another concept similar to chaotic is **ergodicity**. Suppose F is a measurable self-map of a compact topological space K. A Borel measure μ on K is said to be invariant if $\mu(F^{-1}(E)) = \mu(E)$ for all Borel sets $E \subset K$. Suppose that μ is an invariant probability measure on K. We say that μ is ergodic if whenever E is an invariant Borel set, i.e. $F(E) = F^{-1}(E) = E$, then $\mu(E) = 0$ or 1.

Fixed points play a special role in dynamics. Roots p of $R = 0$ are fixed points for the map F from Newton's method. Suppose that F is a holomorphic self-map of a complex manifold M. Suppose that $p \in M$ is a fixed point of F. If all the eigenvalues of $F'(p)$ are strictly less than one, then p is an **attracting** fixed point and there is an open set, the attracting basin of p, consisting of points

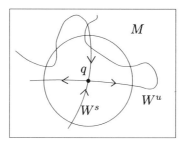

FIGURE 1.6. Stable and unstable manifolds

z for which $F^n(z) \to p$, as we have seen above. If all the eigenvalues are strictly larger than one, the point p is said to be **repelling.** In this case there is an open neighborhood U of p such that if $q \in U \setminus \{p\}$, then there exists an integer $n \geq 1$ so that $F^n(q) \notin U$.

Another important case in higher dimensional dynamics which does not occur in dimension 1 is when some eigenvalues $\{\lambda_i^s\}_{i=1}^k$, but not all, are strictly less than one and all the others, $\{\lambda_j^u\}_{j=1}^l$ are strictly larger than one. Such points are called **saddle** points. Then, there is a neighborhood $U(p)$ of p containing repelling and attracting complex submanifolds $W_{U(p)}^s$, $W_{U(p)}^u$; the **local stable and unstable manifolds** of complex dimension k and l respectively.

The **local** stable manifold $W_{U(p)}^s$ consists of the points $q \in U_p$ so that $F^n(q) \to p$ and $\{F^n(q)\}_{n=1}^\infty \subset U(p)$. The **local** unstable manifold $W_{U(p)}^u$ consists of the points $q \in U(p)$ with a sequence $\{q_n\}_{n=1}^\infty \subset U(p)$, $q_n \to p$, $F(q_{n+1}) = q_n$, $q_1 = q$. See figure 1.6.

The **global stable set,** W_p^s, of p consists of all points $q \in M$ so that $F^n(q) \to p$.

The points considered above, attracting fixes points, repellings points, and saddle points are **hyperbolic fixed points.** In addition there are fixed points where some eigenvalue of the derivative is on the unit circle. The dynamics near these points is much more difficult to describe.

The same terminology applies to **periodic orbit,** $\{z_k\}_{k=0}^l$, $z_{k+1} = F(z_k)$, $z_l = z_0$. They are hyperbolic if $(F^l)'(z_0)$ has no eigenvalue of modulus 1.

The local dynamics near hyperbolic points is **stable,** i.e. maps close to F also have hyperbolic periodic points close to p.

The concept of hyperbolicity generalizes to compact subsets, not only single points or periodic orbits, ([**Ru**]). Let F be a holomorphic self-map on a complex manifold M and let K be a compact subset of M. Assume that K is **surjectively forward invariant**, i.e. $F(K) = K$. We don't assume that F is a homeomorphism, so a point may have several preimages. The set of inverse orbits, $\hat{K} \subset K^{\mathbf{N}}$, $\hat{K} := \{\{p_n\}_{n=-\infty}^{-1} ; F(p_n) = p_{n+1}\}$, is compact in the product topology. We define the tangent bundle, T_K, of $\hat{K}$ as the set of (p, ξ) where $p = \{p_n\} \in \hat{K}$ and where $\xi \in T_M(p_{-1})$ is a tangent vector. Then F lifts to a homeomorphism $\hat{F} : \hat{K} \to \hat{K}$, given by

$$x^2 + x = 1 = 0 \rightarrow x = \frac{-1+i\sqrt{3}}{2}$$

$$x^6 + x + 1 = 0 \rightarrow x =??$$

$$x^3 + xy^2 + 5 = 0 \text{ and}$$

$$y^3 - x^2y + 1 = 0 \rightarrow x, y =??$$

Newton's Method:

$$(x_n, y_n) \rightarrow (x, y)$$

in $\mathbb{C}^2$ or $\mathbb{P}^2$

FIGURE 1.7. Roots of equations

$$\hat{F}(\{\cdots, p_{-2}, p_{-1}\}) = (\{\cdots, p_{-2}, p_{-1}, F(p_{-1})\}).$$

Similarly, F' lifts to a map $\hat{F}'$ on T_K.

We say that F is hyperbolic on K if there exists a continuous splitting $E^u \oplus E^s$ of the tangent bundle of $\hat{K}$ such that the subbundles E^u, E^s are preserved by $\hat{F}'$, and for some constants $C, c > 0$, $\lambda > 1$, $\mu < 1$,

$$\begin{aligned} |(\hat{F}^n)'(\xi)| &\geq c\lambda^n|\xi|, \ \xi \in E^u, \\ |(\hat{F}^n)'(\xi)| &\leq C\mu^n|\xi|, \ \xi \in E^s, \end{aligned}$$

$n = 1, 2, \cdots$.

One of the main open questions for rational maps on $\mathbb{P}^1$ is whether the maps which are hyperbolic on their Julia set are dense in the rational maps.

A weaker property than hyperbolicity is that of stability. A family A of maps $\{F\}_{F \in A}$ is stable at F_0 if there exists an open neighborhood U of F_0 such that all maps $G \in U$ are **topologically conjugate** to F_0, i.e. there exists a homeomorphism $h : M \rightarrow M$ so that $G \circ h = h \circ F_0$.

It is known that the space of rational maps on $\mathbb{P}^1$ is stable on an open dense set ([**MSS**]) when restricted to the Julia set, with the obvious definition of topological conjugacy.

See figures 1.7, 1.8 for an outline of these notes. As is seen from the figures, we do not discuss work on Local Dynamics around one point, Fatou, [**Fa2**], Hakim, [**H**], Ueda, [**U3**] and Weickert, [**W**].

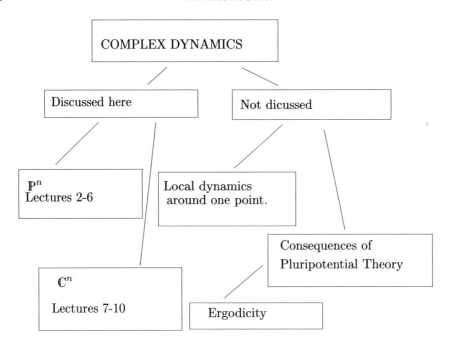

FIGURE 1.8. Outline of Monograph

CHAPTER 2

Kobayashi Hyperbolicity

As mentioned in the introduction, some Fatou components arising in Newtons method are initial guesses leading to roots, but others might not be. Hence, a **main goal is to characterize the nature of the Fatou components**. One of the basic tools in one complex dimension is the Montel Theorem, that is, that the space of holomorphic maps from the unit disc into $\mathbb{P}^1 = \mathbb{C} \cup \{\infty\}$ minus three points is a normal family. Once this tool was discovered, one dimensional complex dynamics got a big push forward because people were then able to prove global results. Before, the theory was mostly local. In several variables, the Kobayashi metric serves this purpose. In this section we first give some basic properties of the Kobayashi metric and then show where the Kobayashi hyperbolicity enters in investigations of iterations of maps.

Let M be a complex manifold; we will mainly deal with open subsets of $\mathbb{C}^n$ and $\mathbb{P}^n$. Pick a point $p \in M$ and a tangent vector ξ to M at p. Let Δ denote the unit disc in the complex plane.

Pick any holomorphic disc through p tangential to ξ, i.e a holomorphic map $f : \Delta \to M$, $f(0) = p$, $f'_*(\partial/\partial z) = c\xi$. The infinitesimal Kobayashi pseudo–metric is obtained by maximizing the discs, $ds(p,\xi) := \inf_f \{\frac{1}{|c|}\}$. See figure 2.1.
As an example, if $M = \mathbb{C}^n$ or $\mathbb{P}^n$, c can be arbitrarily large, so $ds \equiv 0$ in either case. Also, if $M = \Delta$, ds is the Poincare metric.

We say that M is **Kobayashi hyperbolic** if $ds \geq c_K|\xi|$ for some $c_K > 0$, for any compact subset $K \in M$.

Brody ([**La**]) proved a very interesting fact about hyperbolicity on compact complex manifolds.

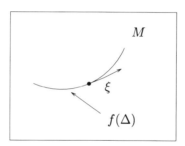

FIGURE 2.1. The Kobayashi metric

THEOREM 2.1. *(Brody) Let M be a compact complex manifold. Suppose that M is not Kobayashi hyperbolic. Then there exists a non constant holomorphic map $f : \mathbb{C} \mapsto M$.*

Pick a smooth Hermitian metric ds on M. Suppose that M is not Kobayashi hyperbolic. Then there exists a sequence of holomorphic maps $\{f_n\}_{n=1}^\infty$, $f_n : \Delta \mapsto M$, $f_n(0) = p_n$, $|f'_n(0)| := ds(p_n, f'_{n_*}(\partial/\partial z)) \mapsto \infty$.

We may assume that f_n is holomorphic in a slightly larger disc.
Define $H_n : \overline{\Delta} \mapsto \mathbb{R}^+$, $H_n(z) = |f'_n(z)|(1-|z^2|)$. Then there exists $z_n \in \Delta$, $H_n(z_n) = \max_\Delta H_n$. Let $g_n : \Delta \mapsto M$, $g_n = f_n \circ \phi_n(w)$ where $\phi_n(w) = \frac{w+z_n}{1+w\bar{z}_n}$. Then

$$|g'_n(w)|(1-|w^2|) = |f'_n(z)||\phi'_n(w)|(1-|w^2|) = |f'_n(z)|(1-|z^2|)$$

(the last equality follows by a short computation).

So now, $|g'_n(w)| \leq \frac{|g'_n(0)|}{1-|w^2|}$. Rescale the disc: Let $R_n := |g'_n(0)|$ (note that $|g'_n(0)| \mapsto \infty$, $n \mapsto \infty$) and define $k_n : \Delta(0, R_n) \mapsto M$, $k_n(z) = g_n(z/R_n)$. Then

$$|k'_n(z)| = \frac{|g'_n(z/R_n)|}{R_n} \leq \frac{|g'_n(0)|}{R_n} \frac{1}{1-|z/R_n|^2} \leq 2$$

say on $\Delta(0, R_n/2)$. Also $|k'_n(0)| = 1$, so using a normal families argument we can find a holomorphic map $f : \mathbb{C} \mapsto M$ such that $|f'(0)| = 1$, hence f is non constant.

Next we mention a few facts which follow directly from the definition of the Kobayashi metric.

THEOREM 2.2. *Holomorphic maps are distance decreasing, i.e., if $F : M \to N$ is a holomorphic map, then*

$$ds(F(p), F'_*(\xi)) \leq ds(p, \xi).$$

Hence invertible maps are isometries. Covering maps are also isometries. Bounded open set in $\mathbb{C}^n$ are Kobayashi hyperbolic

Developing the technique used in the proof of Brody's Theorem, we obtain higher dimensional generalizations of the fact that $\mathbb{P}^1 \setminus$ three points is Kobayashi hyperbolic, which is another way to state Montel's Theorem.
Green ([**Gr1**]) and ([**Gr2**]) proved the following theorem.

THEOREM 2.3. *Let $X_1, .., X_m$ be compact complex hypersurfaces in $\mathbb{P}^n$. Then $\mathbb{P}^n \setminus (\bigcup_j X_j)$ is Kobayashi hyperbolic if*
(i) there is no non–constant holomorphic map from $\mathbb{C}$ to $\mathbb{P}^n \setminus (\bigcup_j X_j)$,
(ii) there is no non–constant holomorphic map from $\mathbb{C}$ to $(X_{i_1} \cap \cdots \cap X_{i_k}) \setminus (X_{j_1} \cup \cdots \cup X_{j_l})$ for any $\{i_1, \ldots, i_k, j_1, \ldots, j_l\} = \{1, \ldots, m\}$.
Assume that the image of a holomorphic map $f : \mathbb{C} \to \mathbb{P}^k$ does not intersect any of $k + 2$ different complex hypersurfaces. Then $f(\mathbb{C})$ is already contained in some compact complex hypersurface.

A holomorphic maps f on $\mathbb{P}^k$ can always be lifted to a homogeneous holomorphic polynomial F on $\mathbb{C}^{k+1}$:

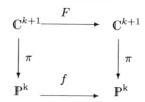

where $\pi : \mathbb{C}^{k+1} \to \mathbb{P}^k$ is the canonical projection. **This F has the important property that it only vanishes at the origin. This fact distinguishes holomorphic maps from rational maps.** Conversely, it is clear that any homogeneous F which vanishes only at the origin gives rise to a holomorphic map on $\mathbb{P}^k$.

In the investigation of dynamics, one always needs a way to count such things as the number of fixed points. For this, the Bezout Theorem is a useful tool.

THEOREM 2.4. *Suppose that $P_1(z_1, ... z_{n+1}), ...,$* $P_n(z_1, ..., z_{n+1})$ *are homogeneous holomorphic polynomials of degree* $d_1, ..., d_n$ *respectively. Their zero sets determine complex hypersurfaces* $X_1, ..., X_n$ *in $\mathbb{P}^n$. If $X_1 \cap X_2 \cap \cdots \cap X_n$ is finite, then the number of points, counted with multiplicity, is $d_1 d_2 ... d_n$.*

As a consequence, we obtain:

THEOREM 2.5. *Let $F : \mathbb{P}^k \to \mathbb{P}^k$ be a holomorphic map of degree d at least two. Then F has $(d^{k+1} - 1)/(d - 1)$ fixed points counted with multiplicity.*

We show next that hyperbolicity occurs generically. Let $\mathcal{H}_d$ denote the holomorphic maps on $\mathbb{P}^2$ of degree d.

THEOREM 2.6. **[FS3]** *Fix an integer $d \geq 2$. Then there exists a dense open set $\mathcal{H}' \subset \mathcal{H}_d$ with the following properties. If $f \in \mathcal{H}'$ and C denotes it's critical set. Then*

i) No point of $\mathbb{P}^2$ lies in $f^n(C)$ for three different n, $0 \leq n \leq 4$.

ii) $\mathbb{P}^2 \setminus \left(\bigcup_{n=0}^{4} f^n(C) \right)$ is Kobayashi hyperbolic.

The proof of the theorem uses analyticity and a calculation near a simple explicit map. The calculation is used to prove the following Lemma.

LEMMA 2.7. *Let $f = [z^d : w^d : t^d]$. There exists an arbitrarily small perturbation g of f such that the five (reducible) varieties $g^n(C)$, $n = 0, \cdots, 4$ have no triple intersections.*

The theorem follows from the Lemma using Greene's Theorem 2.3.

The next result shows that periodic orbits of holomorphic self maps of $\mathbb{P}^k$ are non attracting in the complement of the critical orbits under the hypothesis of Kobayashi hyperbolicity. We say that an open set $\Omega \subset \mathbb{P}^2$ is hyperbolically embedded if Ω is Kobayashi hyperbolic and if in addition the Kobayashi pseudometric of Ω is bounded below by a constant multiple of a smooth metric on $\mathbb{P}^2$.

THEOREM 2.8. *Let $f : \mathbb{P}^k \to \mathbb{P}^k$ be a holomorphic map with critical set C. Let $\mathcal{C}$ be the closure of $\bigcup_{j=0}^{\infty} f^j(C)$. Assume that $\mathbb{P}^k \setminus \mathcal{C}$ is Kobayashi hyperbolic and hyperbolically embedded. If p is a periodic point for f, $f^\ell(p) = p$, with eigenvalues $\lambda_1, \lambda_2, \cdots, \lambda_k$ and $p \notin \mathcal{C}$, then $| \lambda_i | \geq 1$, $1 \leq i \leq k$. Also $| \lambda_1 \cdots \lambda_k | > 1$ or f is an automorphism of the component of $\mathbb{P}^k \setminus \mathcal{C}$ containing p.*

The key idea in the proof is that coverings are isometries in the Kobayashi metric and inclusions are distance decreasing.

DEFINITION 2.9. We say that a Fatou component $\Omega \subset \mathbf{P}^k$ is a Siegel domain if there exists a subsequence f^{n_i} converging to the identity map on Ω.

Using normal families arguments we obtain:

PROPOSITION 2.10. [**FS3**] *Let C denote the critical set of a holomorphic map $f : \mathbf{P}^k \to \mathbf{P}^k$ of degree at least 2. Assume that the complement of the closure of $\bigcup_{n=0}^{\infty} f^{-n}(C)$ is hyperbolically embedded. Then*

$$J \subset \bigcap_{N>0} \bigcup_{n \geq N} f^{-n}(C) =: J(C).$$

Hence all periodic points with one eigenvalue of modulus strictly larger than 1 are in $J(C)$.

THEOREM 2.11. [**FS3**] *Under the assumptions of Theorem 2.8 we have : If there is a component U of the Fatou set of f such that $f^n(U)$ does not converge uniformly on compact sets to C, then U is preperiodic to a Siegel domain Ω with $\partial\Omega \subset C$.*

CHAPTER 3

Some Examples

In this lecture we will give some examples of dynamics in higher dimensions. We first recall a few facts from one variable. There are three types of Fatou components in one dimension:

   i.- Attracting basins
   ii.- Parabolic basins
   iii.- Siegel discs or Herman rings

These have analogues in higher dimension. Let M be a complex manifold and let F be a holomorphic self-map. We have already seen attracting basins.

A fixed point z_0 (or periodic orbit) is called parabolic if at least one eigenvalue has norm one. In this situation, there might be some nonempty open set U, the parabolic basin of z_0 , on which $F^{kn} \to z_0$.

We recall that a Fatou component is a Siegel Domain if there is a subsequence of iterates converging to the identity map. In one dimension, there are two types of such domains, Siegel discs, on which F is conjugate to a rotation, $z \to e^{i\theta}z$ on the unit disc, and Herman rings, on which F is conjugate to $z \to e^{i\theta}z$ on an annulus.

In one variable, these are all the Fatou components, except the preperiodic ones, for which some forward orbit is one of the three above types.

Rational maps F of degree d on $\mathbb{P}^1$ can have at most $2d - 2$ attracting orbits. (See [**CG**, page 58].) This is because there must be at least one critical point in each attracting basin in one variable. Gavosto ([**Ga**]) has shown recently that there are holomorphic maps on $\mathbb{P}^2$ with infinitely many attracting periodic points.

Ueda ([**U1**]) has devised a method to obtain examples of different kinds of Fatou components on $\mathbb{P}^2$.
The map $\Phi : \mathbb{P}^1 \times \mathbb{P}^1 \to \mathbb{P}^2$,

$$\Phi([z_1 : z_2], [w_1 : w_2]) = [z_1w_1 : z_2w_2 : z_1w_2 + z_2w_1]$$

is a 2–1 holomorphic map whose fibers are of the form $\{(z, w), (w, z)\}$. One can push down holomorphic maps on $\mathbb{P}^1$ to holomorphic maps on $\mathbb{P}^2$ using the following commutative diagram:

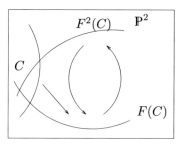

FIGURE 3.1. Critical Orbits

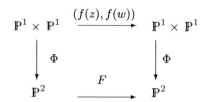

For example, this can be used to find a holomorphic map F on $\mathbf{P}^2$ with a Siegel domain which is of the form of an annulus cross an annulus: Pick a rational map f on $\mathbf{P}^1$ with two Herman rings Ω_1 and Ω_2. Then F has a Siegel domain $\Omega_1 \times \Omega_2$. As another example, one can also use this to find a holomorphic map on $\mathbf{P}^2$ with a dense set of repelling periodic points by picking an f whose Julia set is all of $\mathbf{P}^1$. In particular, this gives an example of a map whose Julia set is all of $\mathbf{P}^2$.

We will next give another, more two dimensional, construction of a map whose Julia set is all of $\mathbf{P}^2$. First let us recall a few facts from the theory of one variable.

The quadratic family $\{P_c(z)\} := \{z^2 + c\}$ consists of degree two polynomials, parametrized by a complex number c. We denote by the Mandelbrot set, M, the set in the parameter plane $\mathbf{C}(c)$ defined by $M := \{c; J_c \text{ is connected}\}$, where J_c is the Julia set of $P_c(z) = z^2 + c$.

It is known that the dynamics is stable on the Julia sets on each interior component of the Mandelbrot set. It is also known that the dynamics on many interior components is hyperbolic. Each such component contains a unique map with a particularly nice dynamics, a so called critically finite map, that is a map for which the critical point has a periodic orbit. The simplest examples are:

1: $c = 0$, $0 \hookrightarrow 0$
2: $c = -1 : z^2 - 1, 0 \mapsto -1 \mapsto 0$

We give a more general definition, valid in two complex dimensions. The definition may be generalized further to arbitrary dimensions.

Let $F : \mathbf{P}^2 \to \mathbf{P}^2$ be a holomorphic map. Let C be the critical set, $C = \{\det R' = 0\}$ with irreducible components $\{C_i\}$. Then, R is **critically finite** if each C_i is periodic or preperiodic. See figure 3.1.

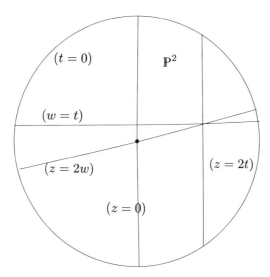

FIGURE 3.2. Critically Finite Map

Furthermore, we say that R is **strictly critically finite** if in addition the 1 dimensional maps $R^{n_i} : R^{l_i}(C_i) \to R^{l_i}(C_i)$ on periodic orbits, are also critically finite maps, i.e. if all their critical points are periodic of preperiodic.

The quadratic maps $P_c(z)$ for which the critical point, 0, is preperiodic, are said to be Misiurewicz points. (See [**CG**, page 133]).)

If one considers polynomials as rational maps on $\mathbb{P}^1$, then one obtains that ∞ is another critical point and it is fixed.

If all critical points of a rational map in one dimension are preperiodic, then the Julia set is all of $\mathbb{P}^1$ ([Mi]).

Our next example ([**FS2**]) shows that there are such maps in two dimension, and that the Julia set is all of $\mathbb{P}^2$.

Let $F : \mathbb{P}^2 \to \mathbb{P}^2$, $F([z : w : t]) = [(z-2w)^2 : (z-2t)^2 : z^2]$. This is the rational map $F : \mathbb{C}^2 \mapsto \mathbb{C}^2$, $F(z, w) = (\frac{(z-2w)^2}{z^2}, \frac{(z-2)^2}{z^2})$ written in homogeneous coordinates.

It is easy to compute the critical set, $C = \{z = 2w\} \cup \{z = 2t\} \cup \{z = 0\}$, and to show that the forward orbit of each component of the critical set ends on the same periodic cycle of three complex lines, $\{z = t\} \mapsto \{w = t\} \mapsto \{z = w\} \mapsto \{z = t\}$. We see that F^3 is a critically finite map on each of these lines. See figure 3.2. All critical points on these lines are strictly preperiodic. Hence by one dimensional theory, the Julia set of F contains these three lines and therefore all their preimages, hence the critical orbit.

The next step is to show that the complement of the critical orbit V is Kobayashi hyperbolic. We see this easily, because it contains a copy of $(\mathbb{C} - \{0, 1\})^2$. The map $F : \mathbb{P}^2 \setminus F^{-1}(V) \mapsto \mathbb{P}^2 \setminus V$ is a covering map, so is an isometry in the Kobayashi metrics for the respective domains. Since $\mathbf{P}^2 \setminus F^{-1}(V)$ is a proper subset of $\mathbf{P}^2 \setminus V$, one can show that the metric is strictly expanding in at least one direction. With this it is possible to show that any Fatou component must converge under iteration to the critical orbit. But this, one can show, leads to a contradiction to the expansiveness of F^3 in the above three lines.

For a rational map on $\mathbf{P}^1$, a finite set E is exceptional if $f^{-1}(E) = E$.

The only exceptional maps on $\mathbf{P}^1$ are $\{z^n\}$, with $E = 0, \infty$.

DEFINITION 3.1. Let $f : \mathbf{P}^2 \to \mathbf{P}^2$ be in $\mathcal{H}_d$ and V a compact subvariety in $\mathbf{P}^2$. Then V is exceptional if $f^{-1}(V) = V$.

We discuss next the exceptional maps in $\mathbf{P}^2$, and give some examples.

PROPOSITION 3.2. [**FS3**] *Let V be the exceptional variety of $f : \mathbf{P}^2 \to \mathbf{P}^2$. Then the degree of V is at most 3.*

To prove the proposition, we write $V = \{h = 0\}$ and show by degree counting that the Böttcher functional equation $h \circ f = f h^d$ holds.

One has the following theorem which describes many examples of exceptional maps.

THEOREM 3.3. [**FS3**] **1.** *If the map $f \in \mathcal{H}_d$ has an exceptional variety containing one complex line, then (we can assume after conjugation with a linear automorphism that) $f = [f_0([z : w : t]) : f_1([z : w : t]) : t^d]$ where the functions $f_0(z, w, 0)$ and $f_1(z, w, 0)$ have nondegenerate degree d -terms.*
2. *If the map $f \in \mathcal{H}_d$ has an exceptional variety containing two complex lines, then we can assume that f has the form $[f_0([z : w : t]) : w^d : t^d]$ where the function $f_0(z, 0, 0) = z^d$.*
3. *If the map $f \in \mathcal{H}_d$ has an exceptional variety which is a union of three lines, then f has the form $[z^d : w^d : t^d]$.*

In addition to the examples in the theorem, the only possible others are if V is a nonsingular quadratic curve $(zw - t^2 = 0)$ and f has odd degree. But no such examples are actually known.

Applications of Kobayashi Hyperbolicity

We investigate Kobayashi hyperbolicity of Fatou components. Our aim is to give further examples of how Kobayashi hyperbolicity can be used to give dynamical results.

At first we discuss some elementary properties of a Green function associated to a holomorphic map.

Let $f = [f_0 : \cdots : f_k]$ be a holomorphic map of degree d in $\mathbb{P}^k$ written in homogeneous coordinates, with lifting $F : \mathbb{C}^{k+1} \mapsto \mathbb{C}^{k+1}$.

We define the function

$$G(z) = lim_{n \to \infty} \log ||F^n(z)||/d^n.$$

This was first introduced by Hubbard in several variables. Brolin ([**Br**]) had introduced it earlier in one complex variable.

See also Hubbard- Papadopol, [**HP**].

THEOREM 4.1. [**FS3**] *G is continuous and plurisubharmonic (except that* $G(0) = -\infty$*).*

Since F vanishes only at 0 we have for some constant M

$$1/M||z^d|| \leq ||F(z)|| \leq M||z||^d$$

and hence

$$1/M||F^n(z)||^d \leq ||F^{n+1}(z)|| \leq M||F^n(z)||^d.$$

Consequently

$$|1/d^{n+1} \log ||F^{n+1}(z)|| - 1/d^n \log ||F^n(z)|||| =$$
$$1/d^{n+1} |\log ||F^{n+1}||/||F^n||^d| \leq 1/d^{n+1} \log M.$$

Consequently G is a continuous (to the extended real line,) plurisubharmonic function in $\mathbb{C}^{k+1}$.

LEMMA 4.2. *G is pluriharmonic precisely on the inverse image of the Fatou set.*

Suppose at first that G is pluriharmonic on some open set $\pi^{-1}(\omega)$. So write $G(z) = \Re H(z)$ for some holomorphic function H such that $1/d^n \log ||F^n|| \mapsto \Re H$. Recall that

$$|1/d^{n+1} \log ||F^{n+1}|| - 1/d^n \log ||F^n||| \leq \log M/d^n$$

so

$$|\Re H - 1/d^n \log ||F^n||| \leq C/d^n.$$

Hence

$$-C \leq \log ||F^n e^{-d^n H}|| \leq C$$

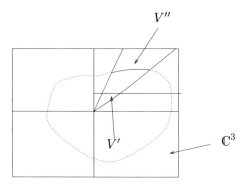

FIGURE 4.1. Kobayashi hyperbolicity of Fatou components

so

$$1/C' \leq ||F^n e^{-d^n H}|| \leq C',$$

hence the iterates are in a normal family.

Next suppose that ω belongs to the Fatou set. So for some subsequence n_k, $f^{n_k} \mapsto \phi(z)$. Hence shrinking ω if necessary, we may assume that we can lift to a sequence $F^{n_k} * \lambda_n(z) \mapsto \Phi(z)$. But then

$$1/d^n \log||F^n|| + 1/d^n \log|\lambda_n| \mapsto \lim 1/d^n \log||\Phi|| = 0.$$

Since the second factor is always pluriharmonic, it follows that G is pluriharmonic.

We prove that Fatou components of holomorphic maps on $\mathbb{P}^2$ are Kobayashi hyperbolic. Note that this rules out that some Fatou component is biholomorphic to $\mathbb{C}^2$ as can be the case for Hénon maps on $\mathbb{C}^2$.

THEOREM 4.3. [**U1**] *Suppose that* $F : \mathbb{P}^k \to \mathbb{P}^k$ *is a holomorphic map of degree* $d \geq 2$. *Then each Fatou component of* F *is Kobayashi hyperbolic.*

We lift F to a homogeneous polynomial map $\tilde{F} : \mathbb{C}^{k+1} \to \mathbb{C}^{k+1}$. Then there is a constant C so that $\frac{1}{C}||z||^d \leq ||\tilde{F}(z)|| \leq C||z||^d$.

Let G denote the Green function. By homogeneity,

$$G(\lambda z) = G(z) + \log|\lambda| \quad (*)$$

It follows that $U = \{z : G(z) < 0\}$ is the basin of attraction of $\{0\}$ for the map $\tilde{F}$ and that the boundary is $\{G = 0\}$. Clearly, U is a bounded set.

We next lift a given Fatou component V to bU. We do this at first locally. Pick a point $p \in V$ and some $q \in \mathbb{C}^{k+1} \setminus \{0\}$, $\pi(q) = p$. Then in some small neighborhood of q, the function G is pluriharmonic and is therefore the real part of a holomorphic function H. Define $\Phi = e^H$. Then by (*), it follows that $\Phi(cz) = |c|\Phi(z)$. Since one of the coordinates of q is non-zero, we can assume say that $q = (q_1, ..., q_k, 1)$ and that we have a set $V' = \{(q_1, ..., q_k, 1)\}$ in this neighborhood, covering a neighborhood of p. Hence, V' is a local lifting of this neighborhood of p.

We choose another lifting V'' which is in the boundary of U by the formula

$$\{\frac{1}{\Phi(q_1, ...q_k, 1)}\}(q_1, ...q_k, 1).$$

We obtain that this lifting is in the boundary of U because $|\Phi| \equiv 1$, hence $G \equiv 0$ there. See Figure 4.1.

Any other local lifting V''' to bU is obtained by multiplying with $e^{i\theta}$, but these are the only liftings to bU. Hence local liftings patch together. Therefore, there is a complex submanifold $\tilde{V}$ in bU which is a covering of V. But U is bounded set, so $\tilde{V}$ and hence V, is Kobayashi hyperbolic.

Next, we give an example how it is easier to work on $\mathbf{P}^k$ using homogeneous coordinates. Also the theory in $\mathbf{C}^k$ is a special case.
We will study normality of inverses of iterates of holomorphic maps $f : \mathbf{P}^2 \mapsto \mathbf{P}^2$ of degree at least two.

Let us first recall the situation in one complex variable.

THEOREM 4.4. *Suppose that $f : \mathbf{P}^1 \mapsto \mathbf{P}^1$ is a holomorphic map of degree $d \geq 2$. Assume there exists an open set $W \subset \mathbf{P}^1$ on which some subsequence f^{j_n} has well defined inverses $g_n := h_{j_n} : W \mapsto U_{j_n} =: W_n$. Then the family of maps g_n is a normal family.*

Note that no W_n can contain a critical point c of f. Moreover, if $j_n > k$, W_n cannot contain any element of $f^{-k}(c)$ either. So if $\cup_{j \geq 0} f^{-j}(C)$ contains at least three points, where C is the set of all critical points, then we are done by Montels Theorem (or the fact that $\mathbf{P}^1 \setminus [0, 1, \infty]$ is Kobayashi hyperbolic). The only case missing is the case of an exceptional map, i. e. a map of the form $f = z^n$ for some non-zero integer n. But for these maps, the result follows by inspection.

This proof generalizes to some cases in higher dimension. Namely, one needs to know that for some intger k, the complement of $\cup_{0 \leq j \leq k} f^{-j}(C)$ is Kobayashi hyperbolic. This is true for a generic family of holomorphic maps, but not for all. And one doesn't have a complete classification of those maps for which this fails in order to handle these maps by inspection. So a better proof is needed to handle the general case. Ueda ([**U2**]) has a proof which generalizes.

[Ueda's Construction for Theorem 4.4:] The map f can be lifted to a holomorphic map $F : \mathbf{C}^2 \mapsto \mathbf{C}^2$ given by homogeneous holomorphic polynomials of degree d. Note that F only vanishes at the origin, so we have the estimates that $\|(z, w)\|^d / C \leq \|F(z, w)\| \leq \|C\| \|(z, w)\|^d$ for some fixed constant C. Hence it follows that there is some constants $0 < r < R < \infty$ such that if $\|(z, w)\| < r$, then $\|F^{-1}(z, w)\| > 2\|(z, w)\|$ and if $\|(z, w)\| > R$, then $\|F^{-1}(z, w)\| < \|(z, w)\|/2$. See figure 4.2. In other words for any point $p \neq 0$ in $\mathbf{C}^2$, and any choice of preimages $F^{-k}(p)$, the sequence satisfies $r < \|F^{-k}(p)\| < R$ for all k large enough.

To prove normality of the map, notice that the map F sends complex lines through zero to complex lines through zero and the map on each line is of the form $t \mapsto ct^d$. Hence the map on each line is unbranched with d possible preimages. If we shrink W, we can find a lifting of f^{-1} to F^{-1} defined on a section of $\mathbf{C}^2$ so that the diagram commutes.

$$
\begin{array}{ccc}
W & \xrightarrow{\ f^{-1}\ } & U_1 \\
{\scriptstyle \pi^{-1}}\downarrow & & \downarrow{\scriptstyle \pi^{-1}} \\
\mathbf{C}^2 & \xrightarrow[\ F^{-1}\ ]{} & \mathbf{C}^2
\end{array}
$$

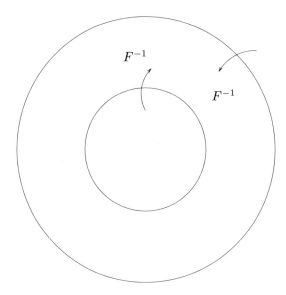

FIGURE 4.2. Normality of Inverse Maps

We can repeat this process to define always inverses F^{-n} on a section of W so that $\pi \circ F^{-n} = f^{-n'} \circ \pi$. But since the images of the F^{-n}'s converge to the bounded set $r < \|(z, w)\| < R$, the maps F^{-n} constitute a normal family. Hence the maps g_n is also a normal family.

Ueda ([**U2**]) proved the same result in general:

THEOREM 4.5. *Suppose that* $f : \mathbb{P}^k \mapsto \mathbb{P}^k$ *is a holomorphic map of degree* $d \geq 2$. *Assume there exists an open set* $W \subset \mathbb{P}^k$ *on which some subsequence* f^{j_n}, *has well defined inverses* $g_n := h_{j_n} : W \mapsto U_{j_n} =: W_n$. *Then the family of maps* g_n *is a normal family.*

The proof is the same in general as in the one dimensional case.

Recall that a Fatou component Ω of a holomorphic map $f : \mathbb{P}^n \mapsto \mathbb{P}^n$ is said to be a Siegel Domain if for some sequence of iterates, f^{n_j} we have that $f^{n_j}|\Omega \mapsto Id$.

In one complex dimension it is known that the boundary of a Siegel disc is contained in the closure of the critical orbit. We will see that as an application of Ueda's result we have the same property in $\mathbb{P}^n$.

THEOREM 4.6. [**U2**] *Let* $f : \mathbb{P}^n \mapsto \mathbb{P}^n$ *be a holomorphic map of degree at least two and suppose that* Ω *is a Siegel Domain. Then the boundary of* Ω *is contained in* $\overline{\cup_{j \geq 0} f^j(C)}$.

REMARK 4.7. One can sharpen the result and show that the boundary is contained in the smaller set $\cap_{k \geq 0} \overline{\cup_{j \geq k} f^j(C)}$.

Fix a point p in the boundary of a Siegel Domain and assume there is a neighborhood W of p which does not intersect the critical orbit. We can assume that W

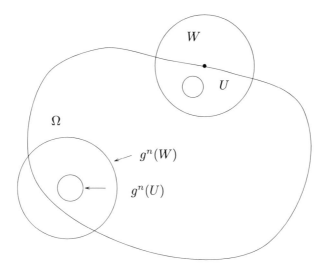

FIGURE 4.3. Siegel Domain

is a ball. Hence we can define for each n a large number of holomorphic inverses of f^n. Pick a nonempty open set $U \subset W \cap \Omega$. Then there is a unique inverse g_n for each n so that $g_n(U) \subset \Omega$. See figure 4.3. For some subsequence $\{m_i\}$, f^{m_i} converges to the identity on U. By the previous Theorem, g_{m_i} converges to the identity on W. Hence f^{m_i} converges to the identity on W. The Theorem now follows from the next Lemma.

Now we use the following Lemma:

LEMMA 4.8. *Suppose* $f : \mathbb{P}^k \mapsto \mathbb{P}^k$ *is a holomorphic map. If* **some** *subsequence* f^{n_j} *converges on an open set* V, *then* V *belongs to the Fatou set.*

Lift f to $F : \mathbb{C}^{k+1} \mapsto \mathbb{C}^{k+1}$ and consider the sequence of plurisubharmonic functions $G_n := \log \|F^{n+1}\|/d^n$. Then this sequence converges uniformly to G. Moreover, the function G is pluriharmonic exactly over the Fatou set. But it follows from the hypothesis that G_{m_i} converges to a pluriharmonic function. Hence V belongs to the Fatou set.

CHAPTER 5

Recurrent Fatou Components

Fix a holomorphic map $f : \mathbb{P}^2 \mapsto \mathbb{P}^2$ of degree $d \geq 2$. We will classify certain periodic Fatou components Ω for f. We can assume that $f(\Omega) = \Omega$ replacing f by an iterate if necessary.

We say that a Fatou component Ω is recurrent if for some $p_0 \in \Omega$ the $\omega-$ limit set of p_0 intersects Ω, i. e. $f^{n_i}(p_0)$ is relatively compact in Ω for some subsequence n_i.

The following theorem has been proved in more generality by Abate under the more restrictive hypothesis that the domains are taut.

THEOREM 5.1. **[FS5]** *Suppose that f is a holomorphic self-map of $\mathbb{P}^2$ of degree $d \geq 2$. Suppose that Ω a fixed, recurrent Fatou component. Then, either :*
i) Ω is an attracting basin of some fixed point in Ω, or
ii) there exists a one dimensional closed complex submanifold Σ of Ω and $f^n(K) \mapsto \Sigma$ for any compact set K in Ω. The Riemann surface Σ is biholomorphic to a disc, a punctured disc or an annulus and $f|\Sigma$ is conjugate to an irrational rotation or
iii) the domain Ω is a Siegel domain.

Since Ω is recurrent, there exists $p_0 \in \Omega$ and n_i such that $f^{n_i}(p_0)$ is relatively compact in Ω. In this case we prove that we are in one of the situations described in i), ii), iii).

Assume $f^{n_i}(p_0) \to p$, $n_{i+1} - n_i \to \infty$. Taking a subsequence $\{i = i(j)\}$ and recalling that we are in the Fatou set, we can suppose that the sequence $\{f^{n_{i(j)+1} - n_{i(j)}}\}_j$ converges uniformly on compact sets in Ω to a holomorphic map $h : \Omega \to \bar{\Omega}$. Let $p_i = f^{n_i}(p_0)$. Then $f^{n_{i(j)+1} - n_{i(j)}}(p_i) = f^{n_{i(j)+1}}(p_0) = p_{i(j)+1}$. Hence $f^{n_{i(j)+1} - n_{i(j)}}(p) = p_{i(j)+1} + O(|\, p_{i(j)} - p\,|)$ so converges to p. Therefore, necessarily $h(p) = p$.

Let $\mathrm{Fix}(h)$ denote the collection of fixed points of h. Since $h \circ f = \lim f^{n_{i(j)+1} - n_{i(j)}} \circ f = f \circ \lim f^{n_{i(j)+1} - n_{i(j)}} = f \circ h$, h commutes with f. It follows that f maps $\mathrm{Fix}(h)$ to itself since if $h(q) = q$, then $h(f(q)) = f(h(q)) = f(q)$.

If, for some h the rank of $h \equiv 0$, then $h(\Omega) = p$, so $p = h(f(p)) = f(h(p)) = f(p)$, hence p is also a fixed point for f. Also both eigenvalues of f' at p must have modulus strictly less than one since some iterates of f converge to the constant map. Hence this leads to case i).

Assume for some h the maximal rank of h is two. Write $h = \lim f^{k_i}$ on Ω. Pick $q \in \Omega$ where h is locally biholomorphic. Say h maps $U(q)$ to V biholomorphically. After shrinking U and choosing i large it follows that $f^{k_i}(\Omega) \supset V$. Hence $V \subset \Omega$. Then, for large i, we have on V that $f^{k_{i+1} - k_i} = f^{k_{i+1}} \circ (f^{k_i})^{-1}$ is approxmately $h \circ h^{-1} = Id$. Hence Ω is a Siegel domain. We are then in case (iii).

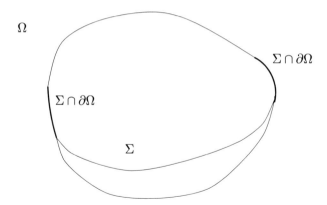

FIGURE 5.1. Recurrent Domains

We now assume that for all h, the maximal rank of h is one. Fix an h and let $\Sigma := h(\Omega)$. Then $\Sigma \subset \bar{\Omega}$. We show first that f is a surjective self map of Σ. Note that Σ is an abstract Riemann surface.

If $x \in \Sigma$ then $x = h(y)$ for some $y \in \Omega$ and $f(x) = f(h(y)) = h(f(y)) \in \Sigma$. So $f(\Sigma) \subset \Sigma$. We show next that the restriction of f to Σ is surjective on Σ.

Let $x = h(y)$, $y \in \Omega$. Choose $y_- \in \Omega$ such that $f(y_-) = y$. Define $x_- = h(y_-)$. Then $x_- \in \Sigma$ and $f(x_-) = f(h(y_-)) = hf(y_-) = h(y) = x$.

Define $\Sigma^0 := \Sigma \cap \Omega$. Since $f(\Omega) \subset \Omega$, then $f(\Sigma^0) \subset \Sigma^0$. Since f is an open map ([FS 1]), f maps the boundary of Ω to itself and hence $f(\Sigma^0) = \Sigma^0$.

After more work, ([FS5]), we obtain that Σ is a disc Δ, Δ^* or an annulus and f acts as rotation $z \mapsto e^{i\theta}z$. The last step of the proof is to show that $\Sigma \subset \Omega$. If Ω was taut, this would have been automatic by definition. (A Kobayashi hyperbolic complex manifold M is **taut** if the family of holomorphic maps from the unit disc to M is a normal family.) But we don't know whether Ω is taut.

To illustrate the idea suppose we have an annulus contained in Σ which in local coordinates is in a complex line $\{(z, 0)\} \subset \mathbb{C}^2$. See figure 5.1

Suppose moreover that $\{1 - \varepsilon < |z| < 1\} \subset \Omega$ and that $\{1 < |z| < 1 + \varepsilon\} \subset \partial\Omega$. Then f has the form

$$f(z, w) = (e^{i\theta}z + wg_1(z, w), wa_1(z) + w^2 k_1(z, w))$$

and

$$f^n(z, w) = (e^{in\theta}z + O(w), w\Pi_{j=0}^{n-1}a_1(e^{ij\theta}z) + O(w^2))$$

For each radius r let $A(r)$ denote the average of $\log|a_1(z)|$ over the circle of radius r. Similarly let $A_n(r)$ denote the average of $\log|a_n(z)|$. Then $A_n(r)$ and $A(r)$ have the same sign always. Also note that the rotation by θ on the circle is ergodic. Hence it follows that $\frac{1}{n}\log|a_n(z)| \to A(r)$ in L^2 on the circle $|z| = r$.

Choose r, such that a_1 does not vanish on the circle $|z| = r$, and hence on a ring $r_1 < |z| < r_2$. Then the functions $\frac{1}{n}\log|a_n|$ are equicontinuous so they converge uniformly to $A(r)$ there.

In particular it follows that if $A(r) < 0$, then the circle with radius r is in the Fatou component. Since $A(|z|)$ is subharmonic, $A(r) > 0$ on the side which belongs to the boundary. But then it follows from ergodicity that for large n, $|a_n(z)| > 1$ uniformly, on circles $|z| = r$. But this implies that these points repell points from Ω. Hence there can be no points in Ω converging to them.

If $f : \mathbf{P}^1 \to \mathbf{P}^1$ is a rational map, there are finitely many recurrent Fatou components. These are either attracting basins, Siegel discs or Herman rings.

There exist on the contrary holomorphic maps on $\mathbf{P}^2$ with infinitely many attracting basins, hence recurrent. ([**Ga**]). This is obtained by using a suitable adaptation to $\mathbf{P}^2$ of the Newhouse phenomenon which produces infinitely many attracting orbits of real diffeomorphisms of compact real two dimensional surfaces.

CHAPTER 6

Rational Maps

In this section we investigate rational self maps in $\mathbb{P}^2$. The main point is to build a skeleton of a theory. So there are mainly definitions in this section, and many obvious research questions.

Notice that if $R = [A : B : C]$ are homogeneous polynomials of degree d, we may assume they have at most finitely many common zeroes (lines of common zeroes in $\mathbb{C}^3$). If not, they have a common factor which can be divided out. The remaining points p if any in $\mathbb{P}^2$ are called points of indeterminacy.

We have that dim $\{A = 0\} = 2$ if A is not equivalent to 0 so gives a complex curve in $\mathbb{P}^2$. Also dim$\{A = B = 0\} \geq 1$ so gives a point in $\mathbb{P}^2$. So usually $\{A = B = C = 0\}$ contains the orgin in $\mathbb{C}^3$ only, hence corresponds to the empty set in $\mathbb{P}^2$. Hence most maps R are holomorphic. Being rational is exceptional.

With the degree d of any homogeneous map $R = [A : B : C]$ we mean the degrees of A, B or C, which are equal, after cancellation of all common irreducible factors. So we assume that $d \geq 2$. Let $I = I(R) = \{q_k\}$ denote the (finite) indeterminacy set consisting of the points q_k in $\mathbb{P}^2$ where $A = B = C = 0$.

For each q_k, the image is a compact Riemann surface S_k, which we call the **blow-up** of q_k. More precisely, $S_k = \cap_{\varepsilon > 0} R(B(q_k, \varepsilon)) \setminus q_k$ where $B(q_k, \varepsilon)$ denotes the ball centered at q_k of radius ε in some arbitrary metric.

Also let $V = \cup V_j$ denote the finite union of irreducible compact complex curves V_j on each of which R has a constant value (at least outside I). Say $R(V_j) = p_j$. We call such curves, V_j, $R-$ **constant**

EXAMPLE 6.1. The Henon map

$$H(z, w) = z^2 + c + aw, z)$$

extend as the map

$$H([z : w : t]) = [z^2 + ct^2 + awt : zt : t^2]$$

on $\mathbb{P}^2$. This map is rational, but not holomorphic. The point $q = [0 : 1 : 0]$ is the point of indeterminacy. The blow-up of q is the line at infinity, $(t = 0)$. The line at infinity is also the $R - constant$ curve for H and its image is the fixed point $[0 : 0 : 1]$.

PROPOSITION 6.2. *If V is nonempty, then also I is nonempty. It can happen that V is empty while I is nonempty.*

Suppose that V is nonempty. Then we may after a rotation assume that some irreducible component $V_j = \{h = 0\}$ for some irreducible homogeneous polynomial h and that $R(V_j) = [0 : 0 : 1]$. Hence h divides both A and B. But then the set $\{C = 0\}$ and $\{h = 0\}$ must intersect and such an intersection point is a point of indeterminacy.

For the converse, consider the example $R = [zw : z^2 + wt : t^2]$. For this example, there is one point of indeterminacy, $[0:1:0]$, while the map is not constant on any curve.

Let R be a rational map $\mathbb{P}^2 \to \mathbb{P}^2$. Let I be the indeterminacy set. Given $a \in \mathbb{P}^2$ we want to discuss the number of preimages of a. Recall that Bezout's Theorem asserts that if $(P_1, \cdots, P_k)$ are k homogeneous polynomials in $\mathbb{P}^k$ with discrete set of zeroes then the number of zeroes counted with multiplicities is equal to the product of the degrees.

Holomorphic maps $\mathbb{P}^2 \to \mathbb{P}^2$ of degree d are $d^2 \mapsto 1$.

PROPOSITION 6.3. *Let $R : \mathbb{P}^2 \to \mathbb{P}^2$ be a rational map of degree d. Assume $I \neq \emptyset$.*
Assume R is of rank 2. Then for any a which is not one of the finitely many points which is the image of an $R-$ constant curve, $R^{-1}(a) = d' < d^2$. Here we count the number of points with multiplicity.

Consider the forward orbit of the points p_j. The $R-$ constant variety V_j is called **degree lowering** if for some (smallest) $n = n_j \geq 0$, $R^{n+1}(V_j) = R^n(p_j) \in I$.

When there is a degree lowering variety, all the components of the iterates of R^{n_j+1} vanish on $V_j = \{h = 0\}$. Hence one need to factor out a power of h in order to describe the map properly. Hence the degree of the iterate will drop below d^{n_j+1}. Hence for rational maps R, the degrees of the iterates R^n can grow erratically. See Medina Bonifant [**MB**] for results in this case.

Next we will study **generic rational maps** on $\mathbb{P}^2$, i.e. rational maps of maximal rank 2, which have degree at least 2 and which have no degree lowering curves. We have first to define Fatou sets and Julia sets of $R : \mathbb{P}^2 \to \mathbb{P}^2$.

LEMMA 6.4. *If I_n denotes the indeterminacy set of R^n, then $I_n \subset I_m \; \forall m > n$.*

The set $I(R)$ should belong naturally to the Julia set because they are blown up. So does $\cup I_n$. Hence the closure $E := \overline{\cup I_n}$, the **extended indeterminacy set**, belongs naturally to the Julia set as well.

PROPOSITION 6.5. *If $p \in \mathbb{P}^2 \backslash E$ and $R^n(p) \in E$, $n \geq 1$, then p is on an R^n- constant curve.*

If p is not on an R^n- constant curve there are arbitrarily small neighborhoods $U(R^n(p))$, $V(p)$ so that $R^n \mid V : V \to U$ is a finite, proper, surjective holomorphic map. Moreover we may assume that $V \cap E = \emptyset$. Since every such open set U contains a point from some I_k, it follows that V contains a point from some I_{n+k}. Hence $p \in E$, a contradiction.

DEFINITION 6.6. Let $R : \mathbb{P}^2 \to \mathbb{P}^2$ be a generic rational map. A point $p \in \mathbb{P}^2$ is in the Fatou set if and only if there exists for every $\epsilon > 0$ some neighborhood $U(p)$ such that diam $R^n(U \backslash I_n) < \epsilon$ for all n.

Note that if ε is small this implies that $I_n \cap U = \emptyset$. Note also that this implies that p cannot belong to the extended indeterminacy set. We say that the Julia set is the complement of the Fatou set.

DEFINITION 6.7. A point $p \in \mathbb{P}^2$ has a **nice orbit** if there is an open neighborhood $U(p)$ and an open neighborhood $V(I)$ so that $R^n(U) \cap V = \emptyset$ for all $n \geq 0$.

So if p has a nice orbit, R^n is well defined for all n on some fixed neighborhood of p. The set of nice points is an open subset of $\mathbb{P}^2 \backslash E$.

DEFINITION 6.8. A generic rational map is said to be **normal** if N, the set of nice points equals $\mathbb{P}^2 \backslash \overline{\cup I_n}$.

Let R be a generic rational map in $\mathbb{P}^2$. With an abuse of notation we will also denote by R a lifting of R to $\mathbb{C}^3$. If $\| \ \|$ is a norm on $\mathbb{C}^3$ we define the nth Green function G_n on $\mathbb{C}^3$ by the formula $G_n := \frac{1}{d^n} \mathrm{Log} \| R^n \|$. Here d is the common degree of the components of R. Observe that if R is rational, G_n has other poles in $\mathbb{C}^3$ than just the origin ($G_n \equiv -\infty$ on $\pi^{-1}(I_n)$). Let π denote the canonical map $\mathbb{C}^3 \mid \{0\} \to \mathbb{P}^2$.

PROPOSITION 6.9. *The functions G_n converge u. c. c. to a function G on the set $\pi^{-1}(N)$ of points with nice orbits.*

If $p \in N$ there exists $U(p)$ and $c > 0$ such that on $\pi^{-1}(U(p))$

$$\| R^{n+1}(z) \| \geq c \| R^n(z) \|^d .$$

On the other hand the reverse inequality

$$\| R^{n+1}(z) \| \leq C \| R^n(z) \|^d$$

holds always. Hence the sequence $\frac{\log \| R^n \|}{d^n}$ converges u. c. c. on $\pi^{-1}(U(p))$.

Because of the second inequality, the limit G always exists and is a plurisubharmonic function on $\mathbb{C}^3$, possibly $\equiv -\infty$, although we don't believe this can happen. (You just need one periodic orbit to show that the limit is not identically $-\infty$.)

THEOREM 6.10. *Let R be a generic rational map on $\mathbb{P}^2$.*
(i) The function G is plurisubharmonic in $\mathbb{C}^3$ (or $\equiv -\infty$).
(ii) G is pluriharmonic on $\pi^{-1}(\Omega)$ if Ω is a Fa̍ ʋu component.
(iii) If N is the set of nice points of R then G is continuous on $\pi^{-1}(N)$ and if G is pluriharmonic on $\pi^{-1}(\omega)$ where ω is an open subset of N, then ω is contained in a Fatou component.

There are two particular cases of generic rational maps which have been previously studied, namely the complex Hénon maps and the class of holomorphic maps on $\mathbb{P}^2$, maps without points of indeterminacy.

We will define various subclasses of the rational maps. These classes are sometimes large enough to contain all holomorphic maps and all Hénon maps.

DEFINITION 6.11. A generic rational map is said to belong to the class of indeterminacy repellors - IR - if there exist arbitrarily small neighborhoods $U \subset\subset V$ of the indeterminacy set for which $R(\mathbb{P}^2 \backslash U) \subset \mathbb{P}^2 \backslash V$.

Both Hénon maps and holomorphic maps belong to IR. Nevertheless the definition is rather strong. It implies that every point in $\mathbb{P}^2 \backslash I$ has a nice orbit, which also implies that the points in I have no preimages. So the extended indeterminacy set E contains I only.

DEFINITION 6.12. We say that a generic rational map belongs to the class with no $R-$ constant blow ups, NRB, if there is no point of indeterminacy q for which the blow up is R^n- constant for some $n \geq 1$. The complement of this class is the set RBwith $R-$ constant blow-ups.

Hénon maps are in RB. The map $R = [zw : z^2 + wt : t^2]$ above is in NRB since it has no R-constant variety.

DEFINITION 6.13. We say that a generic rational map R is a rational Hénon map, RH, if there exists a generic rational map S such that $R \circ S = Id = S \circ R$ in the complement of some hypersurface. We say that S is the inverse of R.

Jeffrey Diller ([**Di**]) has studied these maps and more generally rational maps with rational inverses. He has proved continuity results of the Green's functions both for the map and it's inverse.

CHAPTER 7

Holomorphic Dynamics in $\mathbb{C}^n$

In the previous sections we have mainly been concerned with complex dynamics on complex projective space $\mathbb{P}^2$. The maps we have studied can always be lifted to homogeneous maps on $\mathbb{C}^3$. In the remaining lectures we will discuss iteration of maps on $\mathbb{C}^2$.

We will introduce 5 natural classes of holomorphic maps on $\mathbb{C}^n$. In addition we will discuss flows of holomorphic vector fields. These give rise to mappings which may or may not be everywhere defined. We will consider various natural classes of vector fields as well.

Most of this lecture is used to introduce all these classes of maps and vector fields. In subsequent lectures we will discuss some selected dynamical questions about these. We will show that for dense sets of vector fields dense sets of points explode in finite time (lecture 8). For dense sets of certain maps, orbits are dense (lecture 9) and have dense sets of periodic points (lecture 10).

We let $\mathcal{E}$ denote the collection of holomorphic maps, endomorphisms, on $\mathbb{C}^n$. The set $\mathcal{B} \subset \mathcal{E}$ consists of those maps in $\mathcal{E}$ with inverses in $\mathcal{E}$. The group $\mathcal{B}$ has a subgroup $\mathcal{V}$ of volume preserving maps. On the other hand, if the dimension $n = 2k$ is even, $\mathcal{V}$ has a subgroup $\mathcal{S}$ of symplectic biholomorphisms, i.e. maps which preserve the symplectic form $\Omega = \sum_{i=1}^{k} dz_i \wedge dw_i$. Another subclass of $\mathcal{B}$ consist of the complex Hénon maps $\mathcal{H}$, ([**BS1**], [**BS2**], [**BLS**], [**FS1**], [**Hay**], [**HOV**], [**P**]).

In his thesis, Greg Buzzard ([**B1**]) addressed the question whether the dynamics of biholomorphic maps on $\mathbb{C}^2$ is generically stable. This is one of the main open questions in complex dimension 1. Newhouse ([**N**]) showed that dynamics on two dimensional real surfaces is not generically stable. The key ingredient in his proof was the fact that two sufficiently thich Cantor sets on the real line which intersect, intersect even after small perturbations. This fact fails for Cantor sets in the plane. Buzzard showed however that this stable intersection is still valid for certain Cantor Julia sets. This stable intersection of Julia sets could then be used to produce stable tangencies of stable and unstable foliations, which in turn led to instability of maps.

In the opposite direction, Buzzard ([**B2**]]) showed that for a dense set of maps all periodic points are hyperbolic and all intersections of their stable and unstable manifolds are transverse, i.e. no tangencies. See figure 7.1.

Next we discuss holomorphic vector fields $X = (X_1, ..., X_n)$ where the $X_i's$ are entire holomorphic functions on $\mathbb{C}^n$. In this case, one has for each point $z_0 \in \mathbb{C}^n$ an integral curve $\gamma(t)$ defined for t in some interval $[0, t_0) \subset \mathbb{R}$. This curve then has

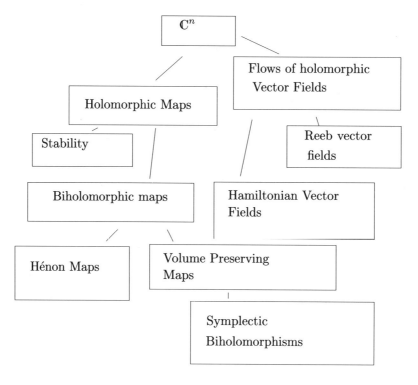

FIGURE 7.1. Dynamics in $\mathbb{C}^n$

the properties that $\gamma(0) = z_0$, and $\gamma'(t) = X(\gamma(t))$.

There is a maximal number t_0, possibly ∞ for which γ is defined. Since the vector field X is holomorphic, one can extend the definition into the complex region, and allow t to be complex. For each z_0, there is a maximal open region R_{z_0} for which γ is defined. In general this maximal region will be a Riemann surface over $\mathbb{C}$. See figure 7.2

Notice that the collection of all curves γ obtained in this way gives a holomorphic foliation of $\mathbb{C}^n$ by complex curves except for points where $X = 0$. These are called **singular points**.
We will in particular consider two kinds of holomorphic vector fields, Hamiltonian vector fields associated to Holomorphic Hamiltonians and Holomorphic Reeb vector fields associated to Complex Contact Structures.

Let us first discuss the Hamiltonian case: Let H be an entire holomorphic function on $\mathbb{C}^{2k}$. Then we define the Hamiltonian vector field

$$X_H := \sum_{j=1}^{k} -\frac{\partial H}{\partial w_j} \frac{\partial}{\partial z_j} + \frac{\partial H}{\partial z_j} \frac{\partial}{\partial w_j}.$$

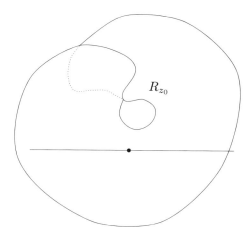

FIGURE 7.2. Domain of Integral Curve

We will also consider the possibility that the Hamiltonian changes with time. In this case the same formula gives rise to time dependent vector fields.

The flow of a holomorphic Hamiltonian vector field preserves Ω. Hence the Jacobian of the flow is identically 1 where defined.

REMARK 7.1. The flow of a Hamiltonian H is along level sets of H.

EXAMPLE 7.2. A shear is a biholomorphism of the following type: $(z, w) \mapsto (z + h(w), w)$ where h is an entire function. Note that shears are given by Hamiltonian flows:
Let $P(w)$ be an entire function. Then the Hamiltonian vector field is $(-P_w, 0)$ and the time 1 map is $(z, w) \mapsto (z - P_w(w), w)$. So we can let $P = -\int h$. A similar remark applies to holomorphic functions $Q(z)$. We obtain shear $(z, w) \mapsto (z, w + Q_z(z))$ as time -1 maps.

The following proposition is well known.

PROPOSITION 7.3. *Let $\Phi \in \mathcal{S}$. Then there exists a C^∞ function $H(z, w, t) : \mathbf{C}^{2k} * \mathbf{R} \mapsto \mathbf{C}$ with period 1 in t such that $H_t(z, w) := H(z, w, t)$ is holomorphic for each t. Moreover Φ is the time 1- map of the time dependent holomorphic Hamiltonian vectorfield $(X_t)_H$.*

Let E denote the class of entire holomorphic functions. We can identify these with the space of **time independent** pluriharmonic Hamiltonians, if we identify entire holomorphic functions differing by a constant.

We recall the concept of a real contact structure in $\mathbf{R}^3$ seen as modelled on the restrictions of a Hamiltonian vector field and a symplectic form on $\mathbf{R}^4$ to a level set of the Hamiltonian. After this we show how the same situation in $\mathbf{R}^8$ leads to complex contact structures on $\mathbf{C}^3 \subset \mathbf{C}^4 = \mathbf{R}^8$.

If we consider the case of a 4-dimensional phase space $\mathbf{R}^4(p_1, p_2, q_1, q_2)$, one can discuss Hamiltonian vector fields, namely, given a Hamiltonian

$H(p_1, p_2, q_1, q_2)$ we consider the Hamiltonian vector field $X = J * \nabla H$ where $J * \partial/\partial p_i = \partial/\partial q_i$ and $J * \partial/\partial q_i = -\partial/\partial p_i$. This gives a vector field X which is tangent to the level sets of H.

Hence we are led to studying vector fields X inside 3 dimensional manifolds S. We can recover the direction up to sign intrinsically from the equation $< \omega, X \wedge Y >= 0$ for all tangential vector fileds Y where ω is the symplectic form on $\mathbf{R}^4$.

Note that $d\lambda|S = \omega|S$, where $\lambda = \sum p_i dq_i$, and one way to determine a length of X is to set $< \lambda, X > \equiv 1$ (or -1 if we want to reverse the vector field).

For the question whether **orbits are singular**, that is, unbounded, what is relevant is the direction of X and not its size. If $S = \{H = 0\}$ and we replace H by gH, we exchange X by gX. Hence intrinsically we need a way to fix a length of X.

However if we are interested in whether **orbits explode** in finite time or not, i. e. the orbits are unbounded on a finite interval, it is more appropriate to use an equation of the form $< \lambda, X > \equiv h$ for a (possibly) variable function. This method breaks down if $\lambda|S$ is zero at some point or if X points in the null space of λ.

To be able to work intrinsically in 3 dimensional surfaces S we start with some $1-$ form λ on S such that $\lambda \wedge d\lambda \neq 0$ and some nonvanishing h. Then (λ, h) is said to be a contact structure and the associated Reeb vector field X is uniquely determined by the equations $< \lambda, X > \equiv h$, $< d\lambda, X \wedge Y > \equiv 0$ for all Y.

Next we will introduce complex odd dimensional contact structures. So these are real even dimensional. Contact structures are usually real odd dimensional. The main point is, however, that for holomorphic Hamiltonians the flow is within complex hypersurfaces rather than real hypersurfaces.
Define

$$z_1 = x_1 + iy_1 = p_1 + ip_2, \; z_2 = x_2 + iy_2 = p_3 + ip_4,$$

$$w_1 = u_1 + iv_1 = q_1 - iq_2, \; w_2 = u_2 + iv_2 = q_3 - iq_4.$$

Let $\Lambda = z_1 dw_1 + z_2 dw_2$ and $\Omega = dz_1 \wedge dw_1 + dz_2 \wedge dw_2$. Then a Hamiltonian H has a holomorphic Hamiltonian vector field if and only if H is pluriharmonic. In this case, $H = \Re F$, F holomorphic and the Hamiltonian vector field is $X = (-\partial F/\partial w_1, -\partial F/\partial w_2, \partial F/\partial z_1, \partial F/\partial z_2)$.

So the flow is within the level sets of F. Next we fix a level set Σ of F, say $F = 0$ and assume $\nabla F \neq 0$. Note that multiplying F by any (invertible) holomorphic function g multiplies the Hamiltonian vector field by g. So we want to recover the direction of X modulo multiplication by complex numbers intrinsically.

LEMMA 7.4. *The Hamiltonian vector field X is the unique solution up to complex multiplication, to the equation $< \Omega, X \wedge Y > \equiv 0$ for all holomorphic vector fields Y tangent to the level set $F = 0$.*

A complex contact form on $\mathbb{C}^3$ is a one form $\Lambda = \sum A_i dz_i$ where the $A_i(z)$ are holomorphic functions and where $\Lambda \wedge d\Lambda \neq 0$.

LEMMA 7.5. *Given Λ and a holomorphic function f, $f \neq 0$, there is a unique holomorphic vectorfield $X = X_f = (a_1, a_2, a_3)$, a_i holomorphic, so that $< d\Lambda, X \wedge Y > \equiv 0$ for all Y and $< \Lambda, X > \equiv f$.*

The vector field $\Re X$ is said to be a holomorphic Reeb vector field.

We need to see if there is an abundance of complex contact forms Λ. Take any locally injective holomorphic map

$$\Phi : \mathbb{C}^3 \mapsto \mathbb{C}^3, \ \Phi = < Z, W, \Gamma > .$$

Then let $\Lambda = dZ + W d\Gamma$.

One has Darboux coordinates locally.

LEMMA 7.6. *Given any complex contact form Λ. Then Λ can be written locally as $dZ + W d\Gamma$ for a local injective holomorphic map (Z, W, Γ). The Reeb vector field becomes $X = (f, 0, 0)$.*

Let X be the Reeb vectorfield, near $p = 0 \in \mathbb{C}^3(z', w', \gamma')$ say, $f \equiv 1$. We can assume $X = \partial/\partial z'$ at zero. Choose new coordinates (z, w, γ) by mapping (z, w, γ) to the point (z', w', γ') which is obtained by integrating X to the (complex) time z starting at $(0, w, \gamma)$. In these coordinates X becomes $\partial/\partial z$. Then Λ has the form

$$dz + A dw + B d\gamma$$

and $d\Lambda = A_z dz \wedge dw + B_z dz \wedge d\gamma + (B_w - A_\gamma) dw \wedge d\gamma$. For $Y = a\partial/\partial z + b\partial/\partial w + c\partial/\partial\gamma$, $< d\Lambda, X \wedge Y > = A_z b + B_z c \equiv 0 \ \forall a, b, c$. So $A_z = B_z = 0$.

It follows that both A and B are independant of z. So $A = \sum a_{i,j} w^i \gamma^j$. Let

$$Z' = z + \sum a_{i,j} w^{i+1} \gamma^j / (i+1), \ W" = w, \ \Gamma' = \gamma.$$

Then

$$dz + A dw + B d\gamma = dZ' - \sum a_{i,j} w^i \gamma^j dw -$$

$$\sum a_{i,j} j w^{i+1} \gamma^{j-1} / (i+1) d\gamma + A dw + B d\gamma = dZ' + B' d\Gamma'$$

Necessarily $\partial B'/\partial w \neq 0$, so use B' as $W'-$ coordinate. Then $p = (0, W_0', 0)$. Finally, write $dZ' + W' d\Gamma' = d(Z' + W_0' \Gamma') + (W' - W_0') d\Gamma'$ and write $Z = Z' + W_0' \Gamma'$, $W = W' - W_0'$ and $\Gamma = \Gamma'$. Then the contact form becomes $dZ + W d\Gamma$ near $p = (0, 0, 0)$.

We finish this lecture by mentioning a result in holomorphic Symplectic Geometry obtained by Franc Forstneric ([**F**]).

The main question is whether a given object can be transformed into another object via a symplectomorphism.

As an example, consider the case of two bounded regions U, $V \in \mathbb{R}^2$ bounded by counterclockwise smooth curves γ_1, γ_2. We ask whether there exists a symplectic diffeomorphism $F : \mathbb{R}^2 \mapsto \mathbb{R}^2$ which maps U onto V. So F preserves the symplectic form $dx \wedge dy$.

An obvious necessary condition is that U and V have the same area. By Green's Theorem this is equivalent to the condition that

$$(*) \qquad \int_{\gamma_1} x \, dy = \int_{\gamma_2} x \, dy.$$

It turns out that $(*)$ is also a sufficient condition for the existence of a symplectomorphism which maps U onto V and hence γ_1 to γ_2.

We will discuss a generalization of this result to the complex case due to Forstneric. First we will give some definitions. Let γ be a simple real analytic smooth closed curve in $\mathbf{C}^{2p}$.

Let $\lambda := \sum_{1 \le j \le p} z_j dw_j$ and let $\Omega := d\lambda = \sum dz_j \wedge dw_j$ be the symplectic form.

DEFINITION 7.7. We call the integral $\int_\gamma \lambda =: A(\gamma)$ the action integral of γ.

Note that if γ_1 and γ_2 are two such curves and if there exists a symplectomorphism Φ on $\mathbf{C}^{2p}$ which maps γ_1 to γ_2, then their action integrals are equal. Namely, write γ_1 as the boundary of a two dimensional surface S_1 and define $S_2 := \Phi(S_1)$. Then

$$A(\gamma_1) = \int_{\gamma_1} \lambda = \int_{S_1} \Omega = \int_{S_1} F^*\Omega = \int_{S_2} \Omega = \int_{\gamma_2} \lambda.$$

Forstneric ([**F**]) proved a version of the reverse.

THEOREM 7.8. *Let γ_1, γ_2 be real analytic smooth simple closed curves in $\mathbf{C}^{2p}$. Assume that $A(\gamma_1) = A(\gamma_2)$. Then there exist neighborhoods U_i of γ_i, $i = 1, 2$ and a symplectomorphism $\Psi : U_1 \mapsto U_2$ such that $\Psi(\gamma_1) = \gamma_2$.*

REMARK 7.9. These maps are not defined globally on $\mathbf{C}^{2p}$. To get a global result one has to introduce extra hypotheses and weaken the statement.
Namely, if one assumes that the curves are polynomially convex, then one can approximate Ψ by a global symplectomorphism $\Phi : \mathbf{C}^{2p} \mapsto \mathbf{C}^{2p}$. In this case the image of γ_1 will be a curve which is close to γ_2, but in general one cannot obtain that the image is exactly equal as in the real case.

REMARK 7.10. The condition that the curves are polynomially convex is automatically fulfilled in the case when the action is non zero. In fact it is a general result by Wermer ([**We**]) that if there exists some holomorphic one form $\sigma := \sum_{1 \le j \le p} A dz_j + B dw_j$ with entire holomorhic coefficients A, B, and the integral $\int_\gamma \sigma \ne 0$, for a simple closed curve γ, then the curve is polynomially convex.

The proof of the theorem goes by finding a time dependent Holomorphic Hamiltonian vector field so that the time 1 map sends γ_1 to γ_2.

CHAPTER 8

Exploding Orbits

In this section we will consider the complex contact structure case in $\mathbb{C}^3$. The integral curves of holomorphic Reeb vector fields lie in Riemann surfaces in $\mathbb{C}^3$ that are not a priori closed.

THEOREM 8.1. [**FG**] *For any complex contact form Λ on $\mathbb{C}^3$, $\exists A \subset C$, C the space of non vanishing entire functions, A a dense $G_\delta-$ set, so that, for every $f \in A$, the interior of K_f, the set of points with bounded orbits, is empty.*

Rather than discussing the proof, we will talk about a refinement of these results. Instead of asking whether orbits are unbounded as time $\mapsto \infty$, we will ask whether orbits explode, i.e. are unbounded on finite time intervals. See figure 8.1.

Let E denote again the space of entire holomorphic functions on $\mathbb{C}^2(z,w)$ with the topology of uniform convergence on compact sets. To each $F \in E$ we associate a holomorphic Hamiltonian vector field $X = (-\partial F/\partial w, \partial F/\partial z)$.

We have the following two results.

THEOREM 8.2. [**FG**] *There is a dense family $G \subset E$ such that every $F \in G$ has a dense set of points with exploding orbits.*

THEOREM 8.3. [**FG**] *Assume that Λ is a fixed complex contact form on $\mathbb{C}^3$. Then there exists a dense family $G \subset C$ so that for every $F \in G$, the contact structure (Λ, F) has a dense set of exploding orbits.*

We discuss at first the proof of Theorem 8.2. Suppose $F \in G$ and that F_n is close to F and has $n \geq 0$ exploding orbits. Pick a point $p \in \mathbb{C}^2$. We want to find F_{n+1} close to F_n with $n + 1$ exploding orbits. We obtain an element $\tilde{F} \in G$ close to F by letting $\tilde{F} = \lim F_n$. See figure 8.2.

First, let us remark that it is sufficient to show that, for any $F \in E$ and any $\beta > 0$, there exists $\tilde{F} \in E$ arbitrarily close to F on $B(0, \beta)$ which has a dense set of points with exploding orbits. To construct $\tilde{F}$, we start with a function $F \in E$

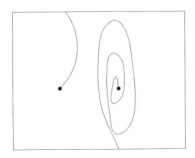

FIGURE 8.1. Exploding Orbits

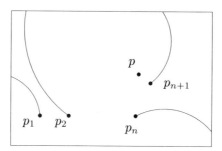

FIGURE 8.2. $n + 1$ exploding orbits

and a dense sequence $\{p_n\} \subset \mathbb{C}^2$. We will make a sequence $\{F_n\}$ of perturbations of F, small on $B(0, \beta)$, so that there exist $\{q_k\}_{k=1}^n$, $\|q_k - p_k\| \leq 1/k$ and the orbit of q_k explodes for F_n and $\{F_n\}$ converges when $n \mapsto \infty$.

Once F_n is chosen, F_{n+1} is chosen so that the orbits of $\{q_k\}_{k=1}^n$ are all unchanged.

We set $F_0 = F$. Suppose F_n has been found. We will define $F_{n+1} = F_n + g_n \prod_{k \leq n}(F_n - F_n(q_k))^2$. The final term will insure that F_{n+1} agrees with F_n to second order on the orbits of q_k, $k \leq n$. We also choose g_n so that $|F_{n+1} - F_n| < 1/2^n$ on $B(0, \beta_n)$ for some sequence $\beta_n \mapsto \infty$.

We will also need to make g_n arbitrarily small on $B(0, \beta)$ to stay in a small neighborhood of F.

If p_{n+1} is not in the Fatou set of F_n then we can find a small perturbation q_{n+1} of p_{n+1} which is not on the level set of any q_k, $k \leq n$, such that the orbit of q_{n+1} is unbounded or a point q_{n+1} in the Fatou set arbitrarily close so we are in the next case.

If p_{n+1} is in the Fatou set, pick at first a small perturbation q_{n+1} on a level set of F_n without critical points. Again assume the point is not on the same level set as any $q_k, k \leq n$.
By rotation one can make the orbit escape a large ball.

Therefore we can assume in all cases that the orbit of q_{n+1} hits $\|z\| = r$ outward, transversally, $r = \max\{\beta, \beta_n\}$. Let Σ be the level set of q_{n+1} for F_n.

We can find a smooth continuation $\gamma(\tau)$ of the orbit so that $\gamma \subset \Sigma$ and $\|\gamma(\tau)\| = \tau$ everywhere, $\tau \geq r$. See figure 8.3

We will modify the orbit to continue in Σ very close to γ and with such a large derivative that it reaches ∞ in finite time.

Choose inductively a sequence of positive numbers $R_m, m \in \mathbb{N}$ satisfying

$$R_0 = r + 1/5, \quad R_{m+1} > R_m \text{ and } R_m \to \infty \text{ as } m \to \infty$$

and so that the projection of $\gamma(\tau)$, $R_{m-1} \leq \tau \leq R_{m+2}$, to the complex line $\mathbb{C}\gamma(R_m)$ is strictly length-increasing. (Set $R_{-1} = r$ say.)

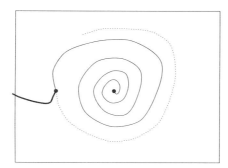

FIGURE 8.3. Changing Orbits

We want to find $g_n = K(F_n - F_n(q_{n+1}))$ where K is a holomorphic function arbitrarily small on $B(0, r)$ and such that the integral curve of some small perturbation of q_{n+1} explodes.

We define K inductively by constructing holomorphic functions k_m, $m \geq 0$, arbitrarily small on $B(0, R_m - 1/5)$ and on a tubular neighborhood of $\gamma[R_m - 1/5, R_{m+2}]$ and such that the integral curve of some small perturbation of q_{n+1} is close to the integral curve of q_{n+1} on $B(0, R_m)$ and goes from $B(0, R_m)$ to $\partial B(0, R_{m+1})$ in arbitrarily small time. Then we just need to continue this inductively to get an exploding orbit.

Let $m \geq 0$. We first outline the steps needed to construct k_m. For simplicity, we will denote k_m by k, R_m by R and assume R_{m+1} and R_{m+2} respectively equal to $R + 1$ and $R + 2$. We may assume $\gamma(\tau)$ is real analytic for $R - 1 < \tau < R + 2$ and is tangent to the integral curve of q_{n+1} to at least fourth order at $\tau = R$.

The first step is to find k_1 real-analytic on $\gamma(\tau)$, $R - 1 \leq \tau \leq R + 2$, so that k_1 vanishes to at least fourth order at $\gamma(R)$;

the Hamiltonian of $F_n + k_1(F_n - F_n(q_{n+1})) \prod_{k \leq n}(F_n - F_n(q_k))^2$ is tangent to $\gamma(\tau)$, $R - 1 \leq \tau \leq R + 2$, and

the corresponding integral curve travels from $B(0, R)$ to $\partial B(0, R + 1)$ as fast as we wish.

To obtain k from k_1, we construct $\tilde{k}$ on Σ, approximating k_1 on $\gamma \cap \{R \leq \|(z, w)\| \leq R + 3/2\}$, and, then, we globalize it to $\mathbb{C}^2$. This will be done in **steps 2 and 3.** In step 2, we use an approximation lemma and in step 3, Hörmander's L^2-estimates for $\bar{\partial}$ with weights.

Step 1

We want a real-analytic function k_1 defined on a neighborhood of $\gamma(\tau)$ in Σ, $R - 1 \leq \tau \leq R + 2$, so that

1- k_1 vanishes to at least fourth order at $\gamma(R)$;

2- the Hamiltonian of $F_n + k_1(F_n - F_n(q_{n+1})) \prod_{k \leq n}(F_n - F_n(q_k))^2$ which is equal on Σ to

$$(1 + k_1 \prod_{k \leq n} (F_n - F_n(q_k))^2) \left(-\frac{\partial F_n}{\partial w}, \frac{\partial F_n}{\partial z} \right)$$

must be tangent to $\gamma(\tau)$, $R - 1 \leq \tau \leq R + 2$, and

3- the corresponding integral curve travels from $B(0, R)$ to $\partial B(0, R+1)$ as fast as we wish.

Step 2

Now, we want to construct $\tilde{k}$. We prove an approximation lemma.

We can assume that the projection $P(L)$ of L to the complex line $\mathbb{C}\gamma(R)$ is a tubular neighborhood ending by arcs of circles and the arc through $P(\gamma(R))$ is tangent to the circle of radius R around zero with opposite concavity.

LEMMA 8.4. *For any $\varepsilon > 0$ there exists a holomorphic function $\tilde{k}_\varepsilon$ on Σ so that*

$$||k_1 - \tilde{k}_\varepsilon||_{\mathcal{C}^1(L)} + ||\tilde{k}_\varepsilon||_{\mathcal{C}^1(\overline{B}(0,R)\cap\Sigma)} + ||\tilde{k}_\varepsilon||_{\mathcal{C}^1(\overline{B}(\gamma(R),\varepsilon)\cap\Sigma)} < \varepsilon.$$

In other words, $\tilde{k}_\varepsilon$ is a good approximation of k_2 in $\mathcal{C}^1$-norm.

Step 3

We want to extend $\tilde{k} = \tilde{k}_\varepsilon$ (ε small enough) in $\mathbb{C}^2$ by solving a $\overline{\partial}$-problem.

Next we discuss the case of complex contact structures.

Recall that a complex contact form on $\mathbb{C}^3$ is a one form $\Lambda = \sum A_i dz_i$ where the $A_i(z)$ are holomorphic functions and where $\Lambda \wedge d\Lambda \neq 0$.

Given $F \in C$, we denote by Φ_t^F the flow associated to X_F.

LEMMA 8.5. *Let Λ be a complex contact form and $F \in C$. The volume form*

$$\frac{\Lambda \wedge d\Lambda}{F}$$

is invariant under the flow Φ_t^F.

Define K_F as the set of points in $\mathbb{C}^3$ with bounded orbits for the flow Φ_t^F. Denote by K the set of $(p, F) \in \mathbb{C}^3 \times C$ so that $p \in K_F$.

Define

$$U_c = \operatorname{int} \left\{ (p, F) \in \mathbb{C}^3 \times C; \quad \sup_{t \geq 0} ||\Phi_t^F(p)|| \leq c \right\}.$$

Then, by a category argument, $U := \cup_{c > 0} U_c$ is dense in $\operatorname{int} K$.

THEOREM 8.6. *For any complex contact form Λ, the interior of K is empty.*

Theorem 8.1 is a corollary of this result.

We discuss briefly the proof of Theorem 8.3.

The contact structure case differs from the symplectic case. One main difference is that after having constructed n exploding orbits in the symplectic case, one can construct the $n + 1^{st}$ without perturbing the previous orbits. In the contact case, any small perturbation of F is likely to destroy the previously constructed exploding orbits. Hence one has to construct them simultaneously.

We start with a dense sequence of points $\{p_n\}_{n \in \mathbf{N}^*}$, $p_n \in B(0, n)$, and with a function $F \in C$. We are going to construct $\tilde{F} \in C$ arbitrarily close to F on $B(0, \beta)$. For simplicity, we will assume that $\beta = 1$.

We make an inductive construction. Namely, we construct a sequence $F_n \in C$ and a sequence of points $\{p_{n,k}\}_{1 \leq k \leq n, \, n \in \mathbf{N}^*}$ so that the following conditions are satisfied:

1- for any $1 \leq k \leq n$, $p_{n,k}$ is close to p_k and $p_{n,k} \in B(0, k)$;

2- F_{n+1} is close to F_n in $\mathcal{C}^0$-norm on $B(0, n)$;

3- the orbits of $\{p_{n,k}\}_{1 \leq k \leq n, \, n \in \mathbf{N}^*}$ relative to F_n goes out of $B(0, n)$ transversally and the time it takes to go from $B(0, k)$ to $\partial B(0, n)$ is at most $\sum_{l=k+1}^{n} \frac{1}{2^l}$. Moreover, the orbit is strictly norm-increasing between $B(0, k)$ and $\partial B(0, n)$.

CHAPTER 9

Unbounded Orbits

We will discuss two similar situations in $\mathbf{R}^n$.

1. Symplectomorphisms F.

2. Hamiltonians F with Hamiltonian vector field X_F and flow Φ.

In either case, we denote by K_F those points in $\mathbf{R}^n$ whose orbits are bounded. The general question going back to Poincaré is whether int K_F is empty for most F. This is open in general, but we will discuss some special cases. See figure 9.1.

THEOREM 9.1. [**FS9**] *There exists a dense G_δ set (S'), $S' \subset S$ so that* int$K_F = \emptyset$ *for all $F \in S'$.*

Roughly speaking, the Theorem says that most orbits go to infinity.

This Theorem confirms a conjecture by Herman

The Theorem isproved for biholomorphic symplectomorphisms on $\mathbf{C}^{2p}$, $p \geq 1$. For simplicity we will discuss here only the $\mathbf{C}^2$ case.

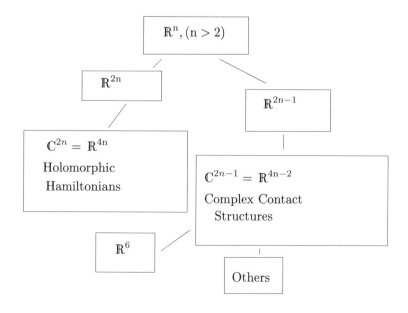

FIGURE 9.1. Poincarés problem

The proof of the Theorem has 4 steps.

Step 1 : Interpret Symplectomorphisms as flows of time dependent holomorphic Hamiltonian vector fields.

Step 2: Prove at first a version of the Theorem in the case of **time independent holomorphic Hamiltonian vector fields**.

Step 3: Prove a version of the Theorem for the space of **time dependent holomorphic Hamiltonian vector fields**.

Step 4: Prove the Theorem.

We discussed Step 1 in section 7.

Step 2

Let us put the topology of uniform convergence on compact sets on E, the space of entire holomorphic function. For each $F \in E$ let K_F denote the set of points with bounded orbit.

For $F \in E$, let

$$U_F := \{(z,w); \exists \, \Omega = \Omega^{open}(z,w) \ \& \ C < \infty;$$

$$\sup_{t \geq 0} \|(z(t), w(t))\| < C \ \ \forall (z(0), w(0)) \in \Omega \}.$$

Notice that by a category argument, U_F is an open dense subset of $\text{int} K_F$.

We call U_F the **Fatou set** of F. This set consists of the largest open set W on which the flow up to time t, Φ_t, is defined for all $0 \leq t < \infty$ and for which $\{\Phi_t\}$ is locally bounded.

Fix a Fatou component U. Then $U = \cup_{C>0} U_C^F$ where $U_C^F = \{(z,w) \in U;$ $\exists \Omega = \Omega^{open}(z,w); \sup_{t\geq 0}\|(z(t), w(t))\| < C \ \ \forall (z,w) \in \Omega \}$.

PROPOSITION 9.2. *For each $t \geq 0$, Φ_t is an automorphism of U_C^F. All limits of $\{\Phi_t\}_{t \in \mathbb{R}}$ are automorphisms of U_C^F. The same conclusion holds for U.*

It is clear from the definition that Φ_t maps each component of U_C into itself. The surjectivity follows from the fact that Φ_t preserves the volume.

Next, consider the orbits of points. There might be (locally) finitely many level sets of the Hamiltonian F for which $\nabla F \equiv 0$. There $\Phi_t \equiv Id$. In addition there might be (locally) finitely many points where $\nabla F = 0$. These are fixed points for the flow.

We have the following result:

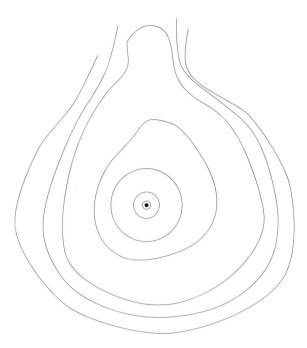

FIGURE 9.2. Integral Curves

THEOREM 9.3. *There exists a dense G_δ set E' of holomorphic Hamiltonians such that $\operatorname{int} K_F$ is empty for all $F \in E'$.*

We will need some preliminary results.

PROPOSITION 9.4. *Let F be a non constant holomorphic Hamiltonian. Let U be a non empty Fatou component. Then each irreducible component of the level sets of F in U is unbounded, and as Riemann surfaces they are isomorphic to Δ, Δ^*, Annulus, $\mathbb{C}$, $\mathbb{C}^*$. Moreover on each of them Φ_t is conjugate to a rotation with the same period.*

To prove this one shows that the closure of $\{\Phi_t\}$ is a T^1 so all orbits lie on circles. These circles are real analytic and foliate level sets of F. If a level set component is bounded one can extend this foliation to a larger set. See figure 9.2.

We need a lemma.

LEMMA 9.5. *Let S be an open Riemann surface. Let $X \not\equiv 0$ be a holomorphic vector field whose orbits are closed. Then for $0 < \theta < \pi$ or $-\pi < \theta < 0$, the orbits of all points, except fixed points, for the vector field $e^{i\theta}X$ spiral to infinity. See figure 9.3*

From the hypothesis on the orbits we know that S is isomorphic to Δ, Δ^*, $\mathbb{C}$, $\mathbb{C}^*$ or an annulus and that the flow is conjugate to the flow obtained by considering the vector field Y on $\mathbb{C}$ given by $Y(z) = icz$ where c is a real constant.

Then $e^{i\theta}X$ is conjugate to the vector field $e^{i\theta}Y$ which satisfies the conclusion.

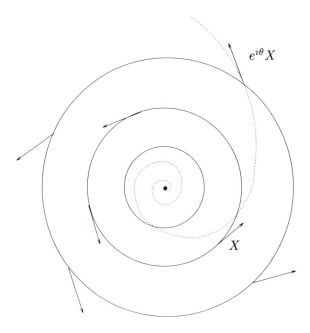

$e^{i\theta}X$

X

FIGURE 9.3. Rotation of a Hamiltonian Vector Field

[Proof of the Theorem] Let $K = \{(z, F); z \in K_F\}$. We show that $\mathrm{int}K = \emptyset$. Suppose not. Then $\exists C > 0$ so that

$$K_C := \{(z_0, F_0); \exists U(z_0, F_0) \text{ such that } z \in U_C^F \;\; \forall(z, F) \in U(z_0, F_0)\}$$

has nonempty interior.

Pick a polydisc $\Delta(z_0; r)$ and a neighborhood $V(F_0)$ so that $\Delta(z_0; r) \times V(F_0) \subset K_C$. Then apply Lemma 9.5 to move an orbit outside $B(0; C)$. Note that if X is the Hamiltonian vector field of F, $X_F = (-F_w, F_z)$, then $e^{i\theta}X$ is thee Hamiltonian vector field of $e^{i\theta}F$. Note also that the level sets of F and $e^{i\theta}F$ are the same. This shows that $\mathrm{int}K = \emptyset$.

Next, pick a basis for the topology of $\mathbb{C}^2$, $\{\Delta_n\}$. For each $F \in E$ there exists an arbitrarily close F' so that some orbit from Δ_n goes outside the ball of radius n. Moreover this is an open condition for F'.

Let $S_n = \{F; \exists p \in \overline{\Delta_n} \text{ so that the orbit of } p \text{ goes outside } \overline{B(0, n)}\}$. Then S_n is open and dense. We define $S' := \cap S_n$. Then for each $F \in S'$, $\mathrm{int}K_F = \emptyset$.

The space E is infinite dimensional. In the case of finite dimensional linear vector sub-spaces of E one can replace G_δ by full measure.

THEOREM 9.6. *Let $F \in E$ be nonconstant. Then for almost all θ, $K_{e^{i\theta}F}$ has Lebesgue measure 0.*

Step 3

To discuss step 3, we will study the class of periodic Hamiltonian flows with period 1, i. e. all smooth $F(t, z, w)$ where $F_t(z, w) := F(t, z, w)$, $F_t \in E$ $\forall t$ and $F_{t+1} = F_t$. Call this space E_t.

We put the topology of uniform convergence in any C^k topology on compact sets on E_t. Hence E_t is a Frechet space.

Fix an $F \in E_t$. Let K_F denote points $(0, z, w) \in \mathbb{R} * \mathbb{C}^2$ for which the orbit $(z(t), w(t))$ is defined for all $t \geq 0$ and

$$\sup_{n \in \mathbb{Z}^+} \|(z(n), w(n))\| < \infty.$$

This does not imply a priori that

$$\sup_{t \geq 0} |(z(t), w(t))| < \infty.$$

Let $U_F := \{(0, z, w) \in int K_F$ so that $\exists C < \infty, \|(z(n), w(n))\| < C$, $\forall (z(0), w(0))$ close to (z, w) and all $n \geq 0\}$.

We consider only time n maps since we are interested in iteration of symplectic automorphisms.

Then U_F is open and dense in $int K_F$. The time 1- map $\Phi : \mathbb{C}^2 = (0) \times \mathbb{C}^2 \mapsto \mathbb{C}^2 = (1) \times \mathbb{C}^2$ is then well defined on U_F and $\Phi(U_F) \subset U_F$. Also Φ is holomorphic and volume preserving.

We have the following result:

THEOREM 9.7. *Consider the product space $H := \mathbb{C}^2 \times E_t$. The set $K \subset H$ consisting of (z, w, F) with bounded orbits of time 1 maps has empty interior.*

The following Corollary is immediate.

COROLLARY 9.8. *There is a dense G_δ of functions F in E_t for which K_F has empty interior.*

We sketch the proof of the Theorem: Suppose K has nonempty interior. For each $C < \infty$ let $K_C = \{(z, w, F) \in int K; \|\Phi_F^n(z, w)\| \leq C \ \forall n \geq 0\}$ where Φ_F is the time 1 map of F.

Fix a $(z_0, w_0, F_0) \in int K_C$. Let U_F denote the connected component of $\Omega_F^C = \{(z, w); (z, w, F) \in int K_C\}$ containing (z_0, w_0) if F is close to F_0. Consider the connected components of Ω_C^F with maximal volume. There can only be finitely many. Then Φ_F' must permute these, so each is periodic. Continuing inductively one sees this is true for all components, also U_F.

We can assume that the period of U_F is constant for F in a neighborhood of F_0. Let $V_F = \cup_n \Phi_F^n(U_F)$.

Then V_F is an open bounded set with $k' < \infty$ components and Φ_F^1 is an automorphism of V_F.

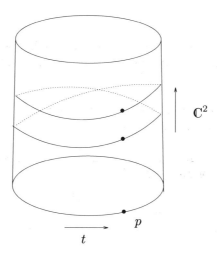

FIGURE 9.4. Periodic Orbit

Define $G_F = \overline{\{\Phi_F^n\}}$. Then $G_F = T^l * A$ where A is a finite commutative group, $l = 0$ or 1. One can exclude the case T^2. The reason is that on a Reinhardt domain in $\mathbb{C}^2$ the automorphisms $(z, w) \mapsto (e^{i\theta}z, e^{i\psi}w)$ only preserve the symplectic form if $\psi = -\theta$. And one can reduce to this situation in local coordinates near a given T^2– orbit.

LEMMA 9.9. *The case $l = 0$ is impossible.*

If $l = 0$, all orbits are periodic with the same period $k := \#A$. Let γ be the closed orbit of some p, $\gamma = (\Phi_F(p, \tau))$ in $\mathbb{C}^3$ where τ runs k times around the unit circle. The projection γ on the unit circle $|\tau| = 1$ is k to 1. See figure 9.4.

We want to construct a smooth function $P(z, w, \tau)$ holomorphic in (z, w) vanishing to second order on γ such that if $F_\epsilon = F + \epsilon P(z, w, \tau)$, $\epsilon > 0$, then p is a hyperbolic periodic point for the flow Φ_τ^ϵ associated to F_ϵ.

Choose a tangent vector ξ and follow it along γ. Then choose for each τ, $P^\tau(z, w)$ so that $(-P_w^\tau, P_z^\tau)$ moves in the direction of ξ away from γ.

The orbit of p remains fixed, but the derivative > 1. So for nearby maps we can not be in U_{F_ϵ}. The reason is that the maps depend holomorphically on ε. Hence for given p, q close, $|\Phi_F^n(p) - \Phi_F^n(q)| \le C|p - q|$ by Schwarz' Lemma.

One next has to rule out $l = 1$ which is more tricky.

The 4th and final step is to study as primary objects the space $\mathcal{S}$ of biholomorphic symplectomorphisms of $\mathbb{C}^2$, not their associated time dependent Hamiltonians.

We give $\mathcal{S}$ the topology of uniform convergence on compact sets. For $f \in S$, let $K_f = \{(z, w); \{f^n(z, w)\}$ is bounded.$\}$

Note that $\mathcal{S}$ contains a dense set C, namely the set of finite compositions of affine linear maps and shears.

The proof of the Theorem 9.1 is analogous to the time dependent case treated before using the fact that symplectomorphisms are time 1- maps of time dependent holomorphic Hamiltonian vector fields.

CHAPTER 10

Generic Density of Orbits

In this section we show that most volume preserving maps have dense orbits. This has as a consequence that for most maps the Fatou set is empty.
At the end of the section we list some open problems.

We denote as before by $\mathcal{V}$ the set of volume preserving biholomorphic maps on $\mathbb{C}^k$, $k \geq 2$.

We prove a version of Theorem 9.1 for volume preserving biholomorphisms.

For $f \in \mathcal{V}$ define as before

$$K_f := \{z; (f^n(z)) \text{ is bounded}\}$$

THEOREM 10.1. [**FS10**] *There is a dense G_δ set $\mathcal{V}_1$ of volume preserving automorphisms so that if $f \in \mathcal{V}_1$, K_f has empty interior.*

We define the product set $K \subset \mathbb{C}^k * \mathcal{V}$ of bounded orbits by

$$K = \{(z, f), \sup_m |f^m(z)| < \infty\}.$$

We are done if we show that K has empty interior. If $U := \text{int } K$ and $C > 0$, let $U_C := \{(z, f) \in U, \sup_m |f^m(z)| \leq C\}$. If K has nonempty interior, then for some C, U_C has also a non empty interior, U_C^0.

Define $\psi : \mathbb{C}^k * \mathcal{V} \mapsto \mathbb{C}^k * \mathcal{V}, \psi(z, f) = (f(z), f)$. Then $\psi(U_C^0) \subset U_C^0$. Since on each slice with fixed f, the map is volume preserving on a bounded region, all components of a slice are periodic.

We fix $(z_0, f_0) \in U_C^0$. The Jacobian determinant, J, of the derivative of f_0 has modulus one. Composing with a rotation which is close to the identity, we can asssume that the J is rational number.

Let Ω_0 be the connected component of the slice $C_{f_0}^k$ which contains z_0. Let Ω be the orbit of Ω_0, and let G be the group generated by f_0. Since G is a commutative compact Lie group, it is isomorphic to $T^l * A$ where A is a finite group and T^l is the torus of dimension l, $l \leq k$, since the orbits must be totally real. If $l = 0$, all points are periodic and we can change one of them into a saddle point by a small volume preserving perturbation ([**B2**]).

We identify G with $T^l * A$ and let G_0 denote the Identity component.

Recall that if F is compact in $\mathbb{C}^k$, the polynomially convex hull $\hat{F}$ of F is defined by

$$\hat{F} := \{z \in \mathbb{C}^k; |P(z)| \leq \sup_{\zeta \in F} |P(\zeta)| \; \forall \text{ holomorphic polynomials } P\}.$$

We want to show that there is a q in the slice of U^0_C over f_0 such it's orbit X_q is polynomially convex.

The point is that holomorphic functions on polynomially convex compact sets can be approximated by global holomorphic functions.

LEMMA 10.2. *Suppose $q_0 \in U^0_C$ and X_{q_0} is not polynomially convex. Then $\exists\, q_1$ so that $\hat{X}_{q_1} \subset \hat{X}_{q_0} \setminus X_{q_0}$. Hence there is a point $q \in \hat{K}_{q_0}$ for which $\hat{X}_q = X_q$. Moreover $\dim X_q \leq k - 1$.*

Pick $q_1 \in \hat{X}_{q_0} \setminus X_{q_0}$. Then the orbits X_{q_0} and X_{q_1} are dijoint. Since $\hat{X}_{q_0}$ is closed under the group action, $X_{q_1} \subset \hat{X}_{q_0}$. Using peak points ([W]), we obtain that $\hat{X}_{q_1}$ and X_{q_0} are disjoint. The final two statements follow from Zorn's Lemma and Serre ([Ho]) respectively.

If $\dim X_q = 0$, q is a periodic point and we can again perturb f_0, ([B2]), to create a saddle point q'. But then q' has an unstable manifold M^u, and by the Schwarz' Lemma M^u must be unbounded. This contradicts that q' is in the interior of K. Hence we can assume that $\dim X_q \geq 1$. Since X_q is totally real and real analytic it has a complexification $\tilde{X}_q$ which is a complex manifold of dimension $1 \leq r \leq k - 1$. The automorphisms in G (we still call the automorphisms on the new slice G and call it still the f_0- slice) preserve $\tilde{X}_q$. In fact, $\tilde{X}_q$ is foliated by orbits of G.

In order to arrive at a contradicition we are going to construct a volume preserving vector field on a neighborhood of $\tilde{X}_q$ pointing away from X_q. Let s be such that $f^s_0 \in G_0$. Let (f_t) be a one-parameter subgroup (f_t) in G_0 with $f^s_0 = f_1$.

The vector field $\xi := \frac{d}{dt} f_t|_{t=0}$ is tangential to each orbit under G_0.

Since we can assume that the Jacobians of all iterates of f^s_0 are rational and since they generate G_0, it follows that all elements in G_0 have Jacobian 1.

Observe that if $\eta = \sum h_j \frac{\partial}{\partial z_j}$ with complex divergence $\sum \frac{\partial h_j}{\partial z_j} \equiv 0$, then it's flow has Jacobian 1. The reverse is also true, hence applies to ξ and $i\xi$.

Hence the flow of $i\xi$ has Jacobian 1 wherever defined.

Next we need to approximate $i\xi$ by a global divergence free vector field. This can be done, using the polynomial convexity of X_q.

There exists hence a divergence free vector field η in $\mathbb{C}^k$ with polynomial coefficients close to $i\xi$ in the component of X_q containing q and close to 0 on the other component.

Then η can be written as a sum of complete divergence free polynomial vector fields (Andersén) and therefore we can approximate the flow of η by a flow ψ_t of volume preserving automorphisms of $\mathbb{C}^k$.

For any $t \in \mathbb{C}$ small enough, $\psi_t \circ f_0$ is close to f_0.
Schwarz's Lemma implies that

$$|(\psi_t \circ f_0)^n - f^n_0| \leq C|t|.$$

This contradicts that ψ_t moves points away from X_q in a transverse direction.

COROLLARY 10.3. *There is a dense G_δ set $\mathcal{V}_2 \subset \mathcal{V}$ with dense orbits.*

To prove the Corollary one can fix any two points p, q in $\mathbb{C}^k$ and any $f \in \mathcal{V}$. After a small perturbation, the iterates of p, q both forward and backward, are unbounded. Then one connects the orbits using a shear, to obtain that p and q are on the same periodic orbit. Next one uses a category argument to produce dense orbits.

We will conclude these lectures with a list of some open problems.

PROBLEM 10.1. Does every holomorphic map $F : \mathbf{P}^k \to \mathbf{P}^k$ of degree at least two have a repelling periodic point?

PROBLEM 10.2. Do rational maps on $\mathbf{P}^2$ always have at least one periodic orbit?

PROBLEM 10.3. Does there exist a holomorphic map $F : \mathbf{P}^k \to \mathbf{P}^k$ with a wandering Fatou component U, i.e. $F^n(U) \bigcap F^m(U) = \emptyset$ for all $n \neq m$?

PROBLEM 10.4. Classify the dynamics around a fixed point of a holomorphic map on $\mathbf{P}^k$ or even just defined in a neighborhood of p.

PROBLEM 10.5. (Herman) Let $F_c : (z,w) \to (z^2 + c - w, z)$, $c \in \mathbf{C}$ a symplectic automorphism of $\mathbf{C}^2$, i.e. F_c is biholomorphic and $F_c^*(dz \wedge dw) = dz \wedge dw$. Can there exist a $c \in \mathbf{C}$ and a periodic orbit $\{z_i\}_{i=0}^k$ for F_c such that z_0 belongs to a Siegel domain ?

PROBLEM 10.6. Let $f : \mathbf{P}^2 \to \mathbf{P}^2$ be a holomorphic map of degree $d \geq 2$. Assume K is a totally invariant set. Let C denote the critical set of f. Assume $\overline{\bigcup_{n=1}^\infty f^n(C)} \bigcap K = \emptyset$. Is f hyperbolic on K?

PROBLEM 10.7. Let H_d denote the space of holomorphic maps $f : \mathbf{P}^2 \to \mathbf{P}^2$ of degree d. This is a finite dimensional space parametrized by the coefficients. Does the set of $f \in H_d$ with infinitely many attracting basins have measure zero?

PROBLEM 10.8. Classify critically finite maps on $\mathbf{P}^3$.

PROBLEM 10.9. (Abate) Are Fatou components for holomorphic maps on $\mathbf{P}^k$ of degree at least 2 always taut?

Index

Bibliography

[AT]   Alexander, H., Taylor, B. A., *Comparison of two capacities in* $\mathbb{C}^n$, Matematische Zeitschrift 186 (1984), 407-417.

[Be]   Beardon, A., *Iteration of rational functions*, Springer Verlag (1991).

[BS1]   Bedford, E., Smillie, J., *Polynomial diffeomorphisms of* $\mathbb{C}^2$, Inv. Math. 87 (1990), 69-99.

[BS2]   Bedford, E., Smillie, J., *Polynomial diffeomorphisms of* $\mathbb{C}^2$ *II*, J. Amer. Math. Soc. 4 (1991), 657-679.

[BLS]   Bedford, E., Lyubich, M., Smillie, J., *Distribution of periodic points of polynomial diffeomorphisms of* $\mathbb{C}^2$, Inv. Math., 114 (1993), 277-288.

[BT1]   Bedford, E., Taylor, B. A., *A new capacity for p. s. h. functions*, Acta Math. 149 (1982), 1-39.

[BT2]   Bedford, E., Taylor, B. A., *The Dirichlet problem for Monge-Ampere equation*, Invent. Math. 37 (1976), 1-44.

[Bi]   Bieberbach, L., *Beispiel zweier ganzer Funktionen zweier Komplexer Variablen, welche eine schlicht volumentreue Abbildung des* $\mathbb{R}_4$ *auf einen Teil seiner selbest vermitteln*, Preuss. Akad. Wiss. Sitzungsber, (1933), 476–479.

[Bie]   Bielefeld, B., *Conformal Dynamics Problem List*, Preprint 1, SUNY Stony Brook, 1990.

[Br]   Brolin, H., *Invariant sets under iteration of rational functions*, Ark. Mat. 6 (1965), 103-144.

[B1]   Buzzard, G., *Ph. D. Thesis, University of Michigan*, (1995)

[B2]   Buzzard, G., *Kupka- Smale theorem for automorphisms of* $\mathbb{C}^n$, preprint.

[CG]   Carleson, L., Gamelin, T., *Complex dynamics*, Springer Verlag, 1993.

[Ce]   Cegrell, U., *Removable singularities for plurisubharmonic functions and related problems*, Proc. London Math Soc. 36 (1978), 310-336.

[CLN]   Chern, S. S., Levine, H., Nirenberg, L., *Intrinsic norms on a complex manifold*, Global analysis. (Papers in honor of K. Kodaira) , Univ. Tokyo Press (1969).

[De]   Demailly J., P., *Courants positifs extrêmaux et conjecture de Hodge*, Invent. Math. 69 (1982), 347-374.

[De]   Devaney, R.L., *An introduction to chaotic dynamical systems*, Addison–Wesley (1989).

[Di]   Diller, J., *Dynamical Green's Functions for Rational and Birational Maps of* $\mathbb{P}^2$, preprint (1995).

[Fa1]   Fatou, P., *Sur les équations fonctionelles*, Bull. Soc. Math. France **47** (1919), 161-271.

[Fa2]   Fatou, P., *Iterations near fixed points*

[FLM]   Freire, A., Lopez, A., Mane, R., *An invariant measure for rational maps*, Bol. Soc. Bras. Mat. 6 (1983), 45-62.

[FS1]   Fornæss, J.E., Sibony, N., *Complex Henon Mappings in* $\mathbb{C}^2$ *and Fatou Bieberbach domains*, Duke Math. J., 65, (1992), 345-380.

[FS2]   Fornæss, J.E., Sibony, N., *Critically Finite Rational Maps on* $\mathbb{P}^2$, Contemporary Mathematics, **137** (1992), 245–260.

[FS3]   Fornæss, J. E., Sibony, N., *Complex Dynamics in higher Dimension I*, Astérisque, 222 (1994), 201-231.

[FS4]   Fornæss, J. E. , Sibony, N., *Complex dynamics in higher dimension. II*. To appear in Ann. Math. Studies.

[FS5]   Fornæss, J. E. , Sibony, N., *Classification of recurrent domains for some holomorphic mappings* , Math. Ann., to appear.

[FS6]   Fornæss, J. E., Sibony, N., *Oka's inequality for currents and applications*, Math. Ann., to appear.

[FS7] Fornæss, J. E., Sibony, N., *Holomorphic Symplectomorphisms in* $\mathbb{C}^2$, to appear in World Scientific Series in Appl. An., *Dynamical Systems and Applications*, 4 (1995).

[FS8] Fornæss, J. E., Sibony, N., *Complex dynamics in Higher Dimensions*, in *Complex Potential Theorey*, ed. Paul M. Gauthier, NATO ASI Series, Math. and Phys. Sci., C439 (1994), 131-186.

[FS9] Fornæss, J. E., Sibony, N., *Holomorphic Symplectomorphisms in* $\mathbb{C}^{2p}$, to appear, Duke Journal.

[FS10] Fornæss, J. E., Sibony, N., *The Closing Lemma for Holomorphic Maps*, preprint (1995)

[FG] Fornæss, J. E., Grellier, S., *Exploding orbits of holomorphic Hamiltonian and complex contact structures*, to appear in Lecture Notes, pure and applied math, Marcel Dekker.

[F] Forstneric, F., *A Theorem in complex Symplectic Geometry*, Journal of Geometric Analysis, to appear.

[Ga] Gavosto, E.A., *Attracting basins in* $\mathbb{P}^2$, Journal of Geometric Analysis, to appear.

[Gr1] Green, M.; *The hyperbolicity of the complement of* $2n + 1$ *hyperplanes in general position in* $\mathbb{P}_n$, *and related results*, Proc. Amer. Math. Soc. **66** (1977), 109–113.

[Gr2] Green, M., *Some Picard theorems for holomorphic maps to algebraic varieties*. Amer. J. Math. **97** (1975), 43–75.

[Gr] Gromov, M., *Entropy, homology and semi-algebraic Geometry*, in Séminaire Bourbaki. Astérisque 145-146 (1987), 225-240.

[Hak] Hakim M., preprint.

[Hay] Hayes, S., *Fatou-Bieberbach Gebiete im* $\mathbb{C}^2$, to appear in DMV-Mitteilungen.

[HaP] Harvey, R., Polking, J., *Extending analytic objects*, Comm. Pure Appl. Math. 28 (1975), 701-727.

[H] Herman, M., Personal Communication, 1993.

[HOV] Hubbard, J., Oberste–Vorth, W., *Hénon mappings in the complex domain I. The global topology of dynamical space*, preprint.

[HP] Hubbard, J. H., Papadopol, P., *Superattractive fixed points in* $\mathbb{C}^n$, preprint.

[Kl] Klimek, M. K., *Pluripotential theory*, Oxford (1991).

[La] Lang, S., *Introduction to Complex Hyperbolic Spaces*, Springer–Verlag, 1987.

[Le] Lelong, P., *Fonctions plurisousharmoniques et formes différentielles positives*, Paris, Londres, New York, Gordon and Breach, Dunod (1968).

[Lyu] Lyubich, M., *Entropy properties of rational endomorphisms of the Riemann sphere*, Ergod. Th. & Dynam. Syst. 3 (1983), 351-385.

[MSS] Mañé,R., Sad P., and Sullivan, D., *On the dynamics of Rational maps* Ann. Sci. Sc. Norm , **16** (1982), 193-217.

[MB] Medina Bonifant, A., *Ph. D. Thesis*, in preparation.

[Mi] Milnor, J., *Dynamics in One Complex Variable: Introductory Lecture*, Institute for Math. Sci., SUNY Stony Brook, 1990.

[N] Newhouse, S., *Diffeomorphisms with infinitely many sinks*, Topology, 13 (1974), 9-18.

[P] Pambuccian, V., preprint.

[deRh] de Rham, G., *Variétés différentiables,* Paris, Hermann (1955).

[Ro1] Rohlin, V. A., *Lectures on entropy theory of measure preserving transformation*, Russian Math. Surveys 22 (1967).

[RR] Rosay, J. P., Rudin, W., *Holomorphic maps from* $\mathbb{C}^n$ *to* $\mathbb{C}^n$, Trans. Amer.Math. Soc., 310 (1988), 47-86.

[Ru] Ruelle, D., *Elements of differentiable Dynamics and bifurcation theory*, Acad. Press, (1989).

[Sc] Schwartz, L., *Théorie des distributions*, Paris, Hermann (1966).

[Sch] Scherbina, *The Levi form for* C^1-*smooth hypersurfaces and the complex structure on boundary of domains of holomorphy*, Izv. Akad. Nauk SSSR 45 (1981), 874-895.

[Sc1] Schröder, E., *Über unendlich viele Algorithmen zur Auflösung der Gleichungen*, Math. Ann. **2**, 1870, 317–365.

[Sc2] Schröder, E., *Über iterierte Functionen*, Math. Ann. **3**, 1871, 296–322.

[Si] Sibony, N., *Unpublished manuscript*, Course at UCLA (1984).

[Sk] Skoda, H., *Prolongement des courants positifs, fermés, de masse finie*, Invent. Math. 66 (1982), 361-376.

[Siu] Siu, Y. T., *Analyticity of sets associated to Lelong numbers and extension of closed positive currents*, Invent. Math. 27 (1974), 53-156.

[S]   Stensönes, B., *Fatou-Bieberbach domain with smooth bondary*, preprint 1995.

[Ta]   Takeuchi, A., *Domaines pseudoconvexes infinies et la métrique riemannicenne dans un espace projectif*, J. Math. Soc. Japan **16** (1964), 159–181.

[T]   Tortrat, P., *Aspects potentialistes de l'itération des polynômes*, Sém. Théorie du Potentiel, Paris, n° 8 (1987). Springer Lecture Notes 1235.

[Ts]   Tsuji, *Potential theory in modern function theory*, Mazuren, Tokyo (1959).

[U1]   Ueda, T., *Fatou set in complex dynamics in projective spaces*, J. Math. Soc. Japan, 46 (1994), 545-555.

[U2]   Ueda, T., *Critical orbits of holomorphic maps on projective spaces*, to appear in Journal of Geometric Analysis.

[U3]   Ueda, T., lectures in Mt. Holyoke, june 1995.

[Wa]   Walters, P., *An introduction to ergodic theory*, Springer Verlag (1981).

[W]   Weickert, B., *Ph. D. Thesis*, in preparation.

[We]   Wermer. J., *Maximum modulus algebras and singularity sets*, Proc. R. Soc. Edinb. 86 (1980), 327-331.

Other Titles in This Series

(*Continued from the front of this publication*)

(See the AMS catalog for earlier titles)